Zero to Einstein in 60™

60 activities and demonstrations guaranteed
to cure science nincompoopitis

© 1989 Rev 1999 • The Wild Goose Company

the
**Wild
Goose**
company
Real Science—Real Fun!®

Greensboro, NC • 888-621-1040

Introduction

Welcome to Wild Goose Science, where education is performance art. Granted, good discipline keeps the hairline from receding, directed lessons are invaluable, and planning helps occasionally, but when it gets right down to it, the time you spend getting those little people excited about something will be the education that they will take with them for a lifetime. As Oscar Wilde said, "Education is what you have left over when you've forgotten everything that you've learned."

Being a highly independent, free-thinking classroom guru, no doubt you can figure out your own uses for the activities in this book. Still, here's a recommendation: There's nothing like a little reward-centered conditioning to help the school day go smoothly, right? You may not get the kids to salivate when you ring a bell, but why not give them an educational treat when they go through a day or week doing — gasp — all that you ask of them? That's when you pull out a **zinger** (one of the cool demos contained herein). The oooh-aaah value of the zingers is high, so you can get away with doing them just for entertainment. With the explanations, though, you can use your favorite zinger to introduce or reinforce all sorts of standard science concepts. Feel free to make copies of the Lab Safety section for your students. It is addressed to them and will serve as a useful reminder that safe is better than sorry.

As mentioned earlier, you have now entered the arena of performance education. Play, make like a kid, draw out the suspense. All of these zingers are fun, very educational, and will be in high demand. Without fail, these experiences are taken home and shared with parents, friends, grandparents, and the Pope when he visits. Children love this stuff. You won't have any discipline problems, their attention spans miraculously double, sometimes even triple, and you'll be their hero. It won't cure warts or get rid of bad breath but *Z to E in 60*™ will give you the tools to be a better science teacher. Above all else, have fun and ham it up.

Without further adieu, here's "Zero to Einstein in 60™" (wild applause).

Table of Contents

Table of Contents

FLUID PRESSURE AND MOVEMENT

Table of Contents

Table of Contents

Table of Contents

Table of Contents

CHEMICAL REACTIONS AND PROPERTIES OF MATTER

Materials

This is a complete list of all the materials that you would need to do every zinger in the book. Find an old orange box and rummage through your kitchen and you'll be 80% of the way there.

acetic acid
air
ball, beach
ball, ping-pong
balls, tennis
balloon, small hot dog-shaped
balloons
banquet fork
battery, 6-V
beaker or clear glass jar
bottle, two-liter
bottle with cap, two-liter
bromothymol blue
broom or wooden dowel
buckets
business cards
calcium chloride
candle
candle, long tapered
cheesecloth
clay, stick
cooking oil
corn syrup, clear
cornstarch
crayon
cup, clear plastic or glass
dime
dish
dowel (about twice as long as the diameter of the Slinky™)
dry ice
eggs, fresh
eggs, hard-boiled
eyedropper
eyedropper, glass
food coloring
food coloring, red
funnel
gas can, unused
glass, tall and clear

gloves, heavy to handle dry ice
guitar
hair
hammer
hands
hot air
hot pads
hot plate
ice
incense stick or cone
index cards
iron filings, (2 ounces)
jar
jar, wide mouth glass
lips and lungs
liquid soap
magnet, bar
magnet, cow
magnets, circular
masking tape, 2" strip
matches
metal object (screwdriver, wrench, etc.)
metal rod, approximately ⅛" to 1" in diameter, and 1' to 3' long
microscope slide
nail, 16-penny
nails
oatmeal cylinder, empty
overhead projector
paper, white
paper muffin cup
paper towel
paper with pattern on it
pencil
petroleum jelly
pie tin, 9"
pipette
plastic bags, 1-quart or smaller
plastic food wrap

polarized filters or two pair of cheap polarized sunglasses
receiver/amplifier plus 2 speakers or a boom box with 2 speakers
rubber band
salt
scissors
sewing spool
Slinky™ or Slinky Jr.™
sodium carbonate
sodium hydroxide solution, 1M
soup can (empty works best)
source of electrons (hair, wool, stray cat)
spoons
steel wool
straight pins
straw with paper wrapper
straws, flexible
straws, plastic
string, ball
tape
test tube (16 mm diameter)
test tube (20 mm diameter)
test tube holder
test tube with cork
thread
tweezers
washer, large
washers
water
water, cold
water, hot
water balloon, homemade
wire, #22 gauge copper
wooden block (approximately 2" x 2" x 4")

Lab Safety

In every lab class there's always the danger that you may expose yourself to injury. The chemicals and equipment you use and the ways you use them are very important, not only for your safety but for the safety of those working around you. Please observe the following rules at all times. Failure to do so increases your risk of accident.

1. Goggles

Goggles should always be worn when chemicals are being heated or mixed. This will protect your eyes from chemicals that spatter or explode. Running water should be available. If you happen to get some chemical in your eye, flush thoroughly with water for **15 minutes.** If irritation develops, contact a physician. Take this book and the bottle of chemical with you the to the doctor's office.

2. Smelling Chemicals

If you need to smell a chemical to identify it, hold it six inches away from your nose and wave your hand over the opening of the container toward your face. This will "waft" some of the fumes toward you without exposing you to a large dose of anything really stinky or dangerous.

3. Chemical Contact With Skin

Your kit contains protective gloves to wear whenever you are handling chemicals. If you do happen to spill a chemical on your skin, flush the area with water for **15 minutes**. If irritation develops, contact a physician. Take this book and the bottle of chemical with you to the doctor.

4. Clean up all Messes Immediately

This is no time to be a pig. Your lab area should be spotless when you start experimenting and spotless when you leave. If not, **clean it**.

5. Proper Disposal of Poisons

If a substance that needs special disposal is used or formed during the experiments in this lab, the book will tell you. These must be handled according to the directions in the lab guide.

Lab Safety

6. No Eating or Drinking During the Lab

When you eat or drink in the lab, you run the risk of internalizing poison. This is never done unless the lab calls for it. Make sure your hands and lab area are always clean when you're finished experimenting.

7. Horseplay Out

Horseplay can lead to chemical spills, accidental fires, broken containers, damaged equipment, and injured people. *Never* throw anything in the lab. Be careful where you put your hands and arms. No wrestling, punching, or shoving!

8. Fire

Remember the rule: No adults in the room = no flames allowed! Get adult help with any fire that's not part of the lab. Know where to locate and how to use a fire extinguisher. If clothes are on fire; STOP, DROP, and ROLL!

9. Better Safe Than Sorry

If you have questions, or if you are not sure how to handle a particular chemical, procedure, or part of an experiment, ask for help from an adult. If you don't feel comfortable doing something, then don't do it. If there is any concern about chemical exposure, contact a physician. Take this book and the bottle of chemical with you.

Lab safety is important! Be safe when you do the experiments in this book or *whenever* you're working in a lab. Have fun!

Characteristics of Water and Other Liquids

Not all liquids are created equal. Water has a molecular structure that causes water molecules to latch on to other water molecules and also on to various materials. Other liquids, such as cooking oil, aren't attracted to water molecules. Still others, such as liquid soap, have molecules that are only halfway attracted. All this attracting and non-attracting leads to some way cool effects and some dazzling demonstrations designed to make the kids think that you're not such a bad teacher after all.

The Floating Egg

Drop an egg into a liquid-filled jar and it sinks. Place the same egg in a second jar and it floats on the surface of the liquid. Hmmmmmm . . .

Materials

2 wide-mouth glass jars   salt
2 fresh eggs   water
1 spoon

Set Up

1. Fill both jars with water.

2. Set one jar aside. This will be where you dunk the egg first.

3. Slowly add salt to jar #2. Stir the water as you pour, making sure that all of the salt dissolves. If you add too much salt or if the water looks milky, don't sweat it; it'll all settle out. See if the egg floats; if not, add more salt (to the water, not the egg). If you still can't get the egg to float, boil the water and add the salt. This will help the salt to dissolve.

The Zing!

1. Place both jars in plain view of the students. Ask them to describe each of the liquids. Have them note the differences and similarities.

2. Gently place the egg on the surface of the water in jar #1 and release it. The egg should tumble to the bottom of the jar. Ask the students to describe what just happened and why it probably happened. Fish the egg out with the spoon.

3. Now, place the same egg on the surface of jar #2 and release it again. It should drop into the water a bit and then bob to the top and stay there. Ask the students to explain why they think this happened. Once again, fish the egg out with the spoon and using the second egg, ask the kids to predict what will happen when this egg is added to the jar and why they think so. It will also float.

4. Remove the egg and pour water out of jar #2 until it is only half-filled with of salt water. Hold a spoon at the surface of the salt water. Gently pour water from jar #1 into jar #2 by pouring directly into the spoon. This will reduce the mixing of the two liquids.

5. Once again, gently place the egg in jar #2. It will sink down until it reaches the salt water in the middle of the jar and then float there in the center of the glass.

The Floating Egg

Howcome, huh?

When the egg (or anything else for that matter) floats, it's due to an upward, supporting force exerted by the fluid called the **buoyant force**. The molecules of water are pushing up on the egg. When you put the egg in the water, it sinks because its weight is greater than the buoyant force of the water inside the jar. The water particles pushing up aren't strong enough to hold up the egg. By adding salt to the water you are smooshing more molecules (particles) into the water. Now the water becomes more dense and there are an increased number of particles pushing up on the egg (greater buoyant force), so it floats.

Extensions

1. Use two eggs instead of one. Place one egg in each solution. The results will be the same, but the students tend to think that the eggs are the cause of the difference.

2. Try other materials that may or may not float. Have the kids construct a data table of all the things they try in the fresh water, and also in the salt water and ask them to record their observations and discuss results.

3. Experiment with different additives. Salt works, but how about sugar, baking soda, baking powder, soap, cornstarch, corn flakes, corn syrup, maple syrup, cooking oil, vinegar, ammonia, or sand? Encourage question asking, recording of data and discussions of results.

4. Challenge the kids to get ten eggs to float at different levels in ten different jars.

5. To show that shape affects buoyancy, give all the students a stick of clay. Ask them to roll it into a ball and drop it into a beaker of water. The clay ball will probably sink. Now ask them to shape the clay so that when it is placed in the water again, it will float. As tempting as it is, try not to give answers. They will eventually figure out that they need to design a boat that will displace enough water so it will float. The more water an object displaces, the greater the buoyant force.

Soap Boat

Chase a paper boat with a drop of soap and it races across water, no wires attached.

Materials

- 1 index card, any size
- 1 pair of scissors
- 1 pie tin, 9", or other container, (must be clean and free of soap)
  liquid soap
  water

Set Up

Fill the pie tin half-full with water and have it sitting on the front table when the students come in. If you are feeling creative name it (e.g. Murphy's Pond), and put a sign up about feeding ducks. This of course has nothing to do with science, but it's fun.

The Zing!

1. Cut a triangle out of the index card (2" high and 1" wide). Hold it up to the students and tell them it's a boat. They probably won't believe you but you've never let this stop you before. Onward into land of scientifica amazamus.

2. Cut a smaller triangle out of the bottom of the boat. Use the drawing at the bottom right as a guide. Tell the students that this is where the motor goes and they'll usually wonder what kind of medication you're on.

3. Place your boat in the water near the edge of the pie tin. It's not going anywhere. This is the appropriate time to put on one of those puzzled looks and then announce that the problem is obvious: there's not any gas in the motor. A petition to have you committed should be circulating by now. Dry-dock your craft (remove the boat from the water).

4. Place a drop of the liquid detergent on your finger and touch it just inside the V-notch of the boat. For those of you who are lost, the "soap here" hint pictured to the right will provide the critical assist. Gently place the boat back in the pan of water, soapy side down, and it will jet across the surface much to the delight and amazement of your students. Wild applause and smiles.

soap here

Soap Boat

Howcome, huh?

All matter is composed of atoms, which contain positively-charged protons, negatively-charged electrons, and neutral (no charge) neutrons. But maybe you already know that. Maybe you also know that atoms clump together to form molecules. So much for Atomic Physics 101. Water molecules look a lot like Mickey Mouse's head. There are two hydrogen atoms (the ears) and one oxygen atom (the head). Atoms of this molecule stick together because they share electrons. While the entire water molecule is neutral (same number of positive and negative charges), the electrons tend to spend more time hanging around the oxygen than they do the hydrogens. This leaves the hydrogen end positively charged and the oxygen end negatively charged. This molecule is called a polar molecule. Sort of like a polar bear that can't decide whether it's coming or going!

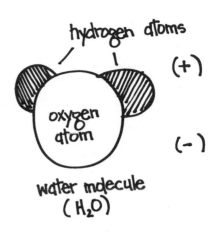

Since opposite charges attract (don't ask why — they just do), polar water molecules tend to cling to each other, with the positive and negative ends attracting. This creates a "skin" on top of the water called **surface tension**.

Enter soap. Soap molecules are very large compared to water molecules. They form long chains. One side of the chain is attracted to water and the rest of the chain isn't, although it is attracted to other soap molecules. When placed on water, the soap molecules tend to break up the water molecules' group hug as they form a thin film of their own. In other words, the soap molecules break up the water's surface tension. The soap film spreads across surface like ripples from a rock dropped in the water, helped along by the fact that the un-soaped water molecules are still clinging to one another.

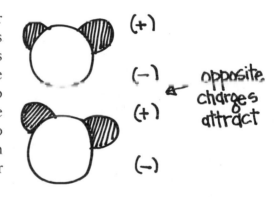

As the soap film spreads, it pushes the paper boat on the edge. The soap boat seems to take off like magic; it's riding the crest of a soap-film wave.

Extensions

1. A chemical called camphor will produce the same effect. Have the students try various chemicals to see if the same effect is produced.

2. This experiment will work on any size surface of water. Try having soap boat races in a large sink or tub.

3. Have the students try different boat shapes. This could lead into a discussion of aerodynamics and friction for older kids, and shapes and their uses for younger kids. The students will come up with some wild ideas.

Soap Boat

Name_____

1. Did the boat move when it was put into the water the first time? Why not?

   _____

   _____

2. What did the boat do when it was put in the water with a drop of soap behind it?

   _____

   _____

3. What did the soap do to the water molecules behind the boat?

   _____

   _____

4. Create three original boat designs, draw them on an index card, cut them out, and test each of them in a container of clean water. Copy the pattern of the fastest boat on to the back of this sheet.

5. Find three other materials that work for making boats and list them here.

   a._____

   b._____

   c._____

6. Design a boat that spins in the water when the soap is added. Once you have mastered that, design a boat that spins the opposite way. Be sure to draw your final designs on the back of this sheet.

Crazy Spinner

Drip some soap on a paper spiral and place it in a pan of water. Watch it spin!

Materials

    1 pair of scissors
    1 pie tin, 9-inch
    1 sheet of paper
       liquid soap
       water

Set Up

1. Fill the pie tin half full of water.

2. Cut two spiral shapes out of the paper. Use the pattern at the right as a guide. The exact thickness of the spiral is of some consequence, but a half-inch is a good start.

The Zing!

1. Ask the kids what they think will happen when the paper is dropped into the water. Once you have enough answers/guesses, place the paper in the water and it should just sit there.

2. This particular spiral is now basically worthless so it can be discarded. Take the second spiral and announce that you are going to add a drop of spinning solution (soap) to the very end and place it in the water. Encourage speculation as to what will happen.

3. Place the design, soapy side down, into the water. This time it should spin like crazy.

Howcome, huh?

Basically, you have the same explanation as for **Soap Boat**. Soap breaks the surface tension, pushes on the spinner, blah, blah, blah.

"OK, fine." you say, "But what makes it spin?" To answer this question, an exhaustive discussion on "rotational dynamics" is needed. On second thought, let's settle for a watered-down, single paragraph.

When the soap was added to the "motor" of the soap boat, the water molecules ended up pushing into the V-notch, propelling it straight forward. If the v-notch was not in the middle of the boat, the boat would tend to go in an arc. Pretty much the same thing is happening here. There is no notch, but because the soap drop is placed where it is, the water molecules push mostly on the square-cut end of the spiral, causing the spinner to rotate. This is best shown by placing a book on the table and giving it a nice even push. It pretty much goes straight forward. Now push it on the corner and see what happens. Behold; spinage, spin-o-rama, nonlinear movement. The same thing is happening with the spinner, except that the water and soap molecules are doing the pushing instead of your hand.

Extensions

1. Spirals of different sizes and thicknesses spin at different rates. Give it a try. Once the water has been "soaped," it must be changed before it can be used again. If the kids don't believe you, let them prove it to themselves.

2. Have the students determine whether or not you have to have the spiral shape, or if the square exposed end is all that is really required. After they figure that out, they can make super spinners with two or three square ends.

3. Once you have mastered square end spinners, try other shapes, such as triangles, semicircles, rectangles, double triangles, hexagons, and dodecahedrons. You invent it; you try it; you explain why you think it did what it did.

4. Encourage the students to try different materials: wax paper, tissue paper, wood, or bottle caps are fun. Create a data table to record the materials and the results.

Crazy Spinner

Name_____

Using the design your teacher taught you as a starting point, experiment with your ideas and invent two spinners; one that spins faster than the design below and one that spins slower than the design below. If you use a material different than paper be sure to identify it on your drawing.

Cheesecloth Cap

Cover a water-filled bottle with cheesecloth. Tip it upside down and the water stays!

Materials
- 1 bottle, two-liter
- 1 rubber band
- 1 square of cheesecloth, 6" x 6" (approx.)
  water

Set Up
No special prep needed.

The Zing!
1. Fill the bottle to the top with water and cover the opening with a square of cheesecloth. Secure the whole operation with the rubber band.

2. Tip the bottle upside down and none (or very little) of the water will come out, even though the cheesecloth is full of holes.

Howcome, huh?
There are a few things going on here. First, the water molecules like each other (see the activity *Soap Boat*) and tend to hold together. On the surface, this is called **surface tension**, but in general, water molecules clinging to one another is called **cohesion**. The water molecules also like the cheesecloth and tend to latch on. This is known as **adhesion**. Inside each little hole of the cheesecloth it's like a whole bunch of critters holding on to each other and onto the fabric of the cheesecloth.

All that holding together still isn't strong enough to overcome gravity, which is pulling the water on through. There must be other mysterious forces at work. And by golly, there are. As the water first starts to drip out of the bottle, it leaves a little air space at the top (formerly the bottom). Since no air is able to enter the bottle from below, this air space at the top contains relatively few air molecules. With fewer air molecules to push on the water at that point, you have an area of low pressure. The air pressure outside of the bottle pushing up on the water at the opening is much higher. This difference in air pressure pushes up on the water and helps keep it in the bottle. For more information on fluid pressure, check out that section of the book.

Cheesecloth Cap

Extensions

1. Experiment with different-sized openings and different containers. You can ask the students to bring in any of a number of containers and really give this some lab time.

2. Experiment with different fabrics and chart the effectiveness of the fabric compared with the size of the holes in the fabric.

Cheesecloth Cap

Name_____

List several materials that you think will prevent water from falling out of the bottle, like the cheesecloth. List the materials and then test each of them. Record the results of your tests and your comments in the spaces below.

Material	Result
1._____	_____
2._____	_____
3._____	_____
4._____	_____
5._____	_____
6._____	_____
7._____	_____
8._____	_____
9._____	_____
10._____	_____
11._____	_____
12._____	_____
13._____	_____
14._____	_____
15._____	_____

Pinching Water

Pinch independent streams of water into one connected stream; then, separate them with the flick of a finger.

Materials
> 1 hammer
> 1 nail
> 1 soup can (empty works best)
> masking tape, 2" strip
> water

Set Up

1. Using the nail and hammer, punch two holes near the bottom of the soup can. They should be about a half-inch from the bottom of the can and one quarter-inch apart. You can experiment with different hole placements and find a better set-up, but this one should work just fine for the no-nonsense experimenters in the crowd.

2. Place the masking tape over the holes, making sure you have a good seal, and fill the can with water.

The Zing!

1. Over a sink or bucket, or outside, hold the can up so all of the students can see it and remove the tape. Two distinct streams of water should shoot out of the can.

2. Announce that both streams of water have a secret longing for one another and if they could just get close enough they would be inseparable. The question of your personal sanity may now arise, but proceed anyway. Using your thumb and index finger, pinch the water together into a single stream. A word of advice to the rookie science teacher: the key to good demonstrations is practice prior to performance.

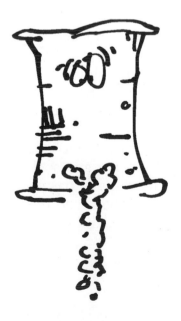

3. The streams of water will stay joined until you flick them using your index finger. They should separate again. If you work the "love gone sour" angle, it can be pretty funny.

Pinching Water

Howcome, huh?

Water molecules are polar and are attracted to one another. This is called **cohesion** (see previous activities). Once you pinch the streams together, the molecules in the separate streams tend to stay together. There's a bit of the Bernoulli Effect (see the fluid pressure section) going on here, too. After you do the activities in that section, come back to this one and see if you can figure out how the Bernoulli Effect might apply.

When you flick the water streams with your finger, you're simply using the force of your finger to overcome the cohesion of the water streams. Makes you feel powerful, huh?

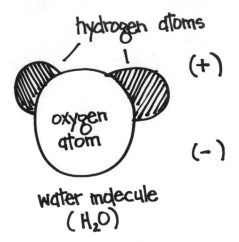

hydrogen atoms

(+)

oxygen atom

(−)

water molecule (H₂O)

Extensions

1. It's fun to experiment to see how close the streams have to be before they automatically come together; but this can only be done if there are a whole bunch of soup cans hanging around. Once the kids have figured that out, have them see how far apart the streams of water can get before they remain separated.

2. Show that the volume of water has little to no effect on the results.

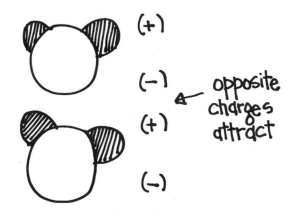

(+)

(−)

(+)   opposite charges attract

(−)

3. See if the demonstration can be repeated using three holes punched in a line. How about if they're punched in a triangle? Does it really matter? If you can do three holes, how about four or five? How many holes can you punch and what shapes are most conducive to achieving the desired end?

4. Take two soup cans with no holes in them and fill one with water. Dunk some cotton string into the water and soak it thoroughly. Stretch the string between the two cans and pour water from the can that has water, into the can that does not. Do <u>not</u> do this over the table. The adhesion of the water to the string allows you to transfer the water from can to can without spilling the proverbial drop.

Pinching Water

Name_____

1. What happened when the water was pinched?

   _____

   _____

2. How did your teacher unpinch the water?

   _____

   _____

3. Mix some salt into the can and repeat the activity. Do you get the same results? Why?

   _____

   _____

   _____

4. Devise an experiment that demonstrates this same principle a different way. Outline your experiment proposal on a separate sheet of paper and ask your teacher to review it before you begin. Be sure that you include the following components:

   Hypothesis

   List of Materials

   Procedure

   Data and Results

   Conclusion

Mixing Oil and Water

Shake a test tube, mixing oil and water. Let the oil and water settle and what happens? Now, add a drop of soap and change the results.

Materials

1 test tube with cork
cooking oil
liquid soap
water

Set Up

No special prep needed.

The Zing!

1. Fill one-third of the test tube with water. Add another one-third of oil and stopper the test tube. Note which liquid floats on top.

2. Shake this mixture thoroughly and then have a student volunteer hold the tube while the suspension settles out. The oil and water will separate back into their respective layers.

3. Add a couple of drops of soap to the tube and shake it again. This time the two solutions will not separate.

Howcome, huh?

For starters, think about why the oil floats on top of the water instead of mixing in. In order to understand the explanation, you have to know something about **density**. The density of something is, technically, the mass of the something divided by its volume. In everyday terms, it goes like this: A bunch of feathers loosely packed in a box have a smaller density than that same bunch of feathers crumpled and smashed into a small ball. A box full of metal ball bearings is more dense than that same box full of cotton balls. It depends on how much mass (stuff) you have in a given space.

Liquids that are more dense tend to sink down and displace liquids that are less dense. So cooking oil is obviously less dense than water, because the cooking oil floats on top of the water.

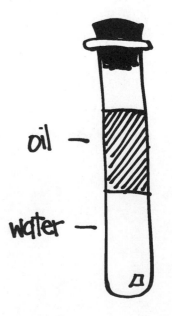

Mixing Oil and Water

If you've ever tasted the two, you also know that cooking oil is different from water in other ways; taste, for example. As you know from previous zingers, water is a polar molecule, meaning that it's positive on one side and negative on the other. Because of electrical attractions, polar molecules are attracted to other polar molecules. Oil is a non-polar molecule. Non-polar molecules are attracted to each other (through a mechanism known as electron sharing), and they're not attracted to polar molecules. In other words, oil and water don't mix! The less dense oil floats on top, while the dense water sinks to the bottom. Shake 'em up and it doesn't take long for the molecules to return to their original positions.

Soap molecules are a bit different. They have a polar side and a non-polar side. One side is attracted to water molecules and the other side is attracted to oil molecules. The soap acts as a bridge, bringing the water and oil molecules together. With all three molecules hooked together, there's very little separation, once you shake them up. The differences in density aren't enough to keep the two liquids separate. This makes soap a very good thing for getting oils, food, and dirt out of clothes and off of hands.

Extensions

You can observe the same reaction using vinegar and oil. Looks like it's time to make a nutritious salad and then add oil and vinegar (no soap!) dressing.

Mixing Oil and Water

Name_____

1. Why do you think the oil and water separated after the test tube was shaken?

    _____

    _____

2. What happened when soap was added to the mixture?

    _____

    _____

3. Why do you think soap is useful in cleaning hands?

    _____

    _____

4. Draw pictures of each test tube, the first without soap and the second with soap.

Cartesian Diver

Float an eyedropper in a water-filled plastic bottle. Replace the cap, squeeze hard, and the eyedropper sinks! Release the pressure and it floats again.

Materials
　　1 bottle with cap, two-liter
　　1 glass, tall and clear
　　1 glass eyedropper
　　　water

Set Up
Fill the bottle almost all the way to the top with water. Draw up as much water as you can into the eye dropper and plop it, bulb-side-up, into the bottle. The dropper should float. If not, start over, drawing a little less water into the dropper. Cap the bottle tightly and squeeze the sides. The dropper should sink to the bottom of the bottle. If the dropper doesn't sink, no matter how hard you squeeze, draw more water into the dropper and try again.

The Zing!

1. Place the eyedropper in the bottle and ask the students to observe your demonstration. Tell them that there is an invisible treasure at the bottom of the bottle and you want them to figure out a way for the diver (eyedropper) to get down there to retrieve it.

2. After the suggestions have rolled in, place both hands on either side of the bottle and squeeze. The diver should sink to the bottom of the jug. Release, and the diver should head to the top. With practice, you can control the speed at which the diver rises and sinks.

3. Ask the kids to explain what is going on. With close observation, they should notice that the water level inside the dropper changes as you squeeze and release.

Cartesian Diver

Howcome, huh?

When you placed your hands on the bottle and squeezed, the pressure inside the jug increased. You were squishing the water molecules together and they began to look for a place to go to release that pressure. Since liquids do not compress as easily as gases, the air in the eyedropper was mushed together and more water went into the eyedropper. The added weight of the water inside the eyedropper caused it to sink. When the pressure was released, the water came out and the eyedropper was buoyant enough to rise to the surface again.

This same principle is used in the operation of submarines. Water is allowed to enter large tanks when the submarine dives. To surface, air is forced into the tanks to expel the water.

Extensions

1. Try using other solutions. Sugar water and salt water work well. You'll get different results depending on the density of the liquid.

2. Experiment with different levels of water in the dropper and notice how hard you have to squeeze to get the dropper to sink. If you're feeling ambitious, you could graph the water level in the dropper against the strength of squeeze (measured maybe, by how far the bottle is pushed in on the sides).

Cartesian Diver

Name_____

1. What happened when the diver was placed in the bottle and pressure was applied to the sides?

   _____

   _____

2. What happened when the pressure was released?

   _____

   _____

3. When pressure was applied to the bottle, the air inside the eyedropper was squished together. This makes the eyedropper less buoyant (it won't float as well). Does this explain why the eyedropper sank to the bottom? Explain.

   _____

   _____

   _____

4. How can you control the depth to which the eyedropper sinks?

   _____

   _____

   _____

   _____

   _____

Paper Worms

Squish a straw wrapper into a small accordion. Add a few drops of water and the straw turns into a worm. Well, almost!

Materials

1 straw with paper wrapper
pipette or eyedropper (if available)
water

Set Up

Make sure your straw is squished up really tight.

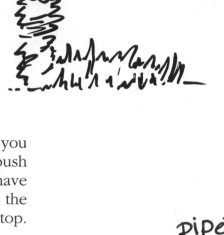

The Zing!

1. Scrunch the wrapper into the middle from both ends. If you have trouble doing this, lick the ends of your fingers and push from the ends of the straw in to the middle. You will have something that looks like the illustration above. Remove the smooshed wrapper from the straw and lay it on the table top.

2. Using a pipette, eyedropper, or the straw and your finger, add a couple of drops of water to the wrapper. It will begin to move outward. Don't scream, this sets a bad precedent for the children. These are non-biodegradable worms, so don't throw them outside for the birds to eat.

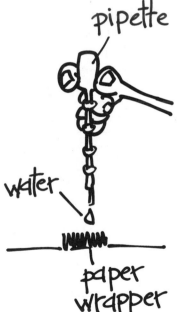

Howcome, huh?

The water molecules are so attracted to the paper (adhesion), they get completely absorbed. As the water heads into the fibers of the paper, it makes them fatter and longer, causing the many folds in the wrapper to push apart. Instant worm.

Extensions

1. Have the students use different liquids to see if it's the water that creates the commotion or if it's just liquid in general.

2. Get a real worm and have a race to see who is faster. Just kidding.

32

Paper Worms

Name_____

Outline this activity so one of your friends could do it if they wanted. Include the materials, the procedure, a place to record data, and what you think happened.

Materials

_____

_____

Procedure

_____

_____

_____

_____

Data

_____

_____

_____

Conclusion

_____

_____

_____

_____

_____

34

Fluid Pressure and Movement

Scientists consider liquids (like water) and gases (like air), to be fluids. That's because liquids and gases behave almost identically under pressure and when they flow over surfaces, as fluids are apt to do. One big difference though, is that it's easy to compress gases and liquids are virtually incompressible. At any rate, most of the zingers in this section are about air pressure. Information about liquid pressure and movement is included at the end.

Wrinkled Can

Fill an empty, never-used gas can with a cup of water and heat it up. When the water boils, remove the can from the heat and cap it tightly. The can collapses with cool crinkling sounds.

Materials

1 beach ball

1 cup of water

1 hot plate

1 empty, unused gas can

hot pads

Set Up

1. Locate an clean, unused gas can.

2. Plug in the hot plate ahead of time so you don't have to wait for it to heat up.

3. Get the cup of water from the sink.

The Zing!

1. Pass the unused gas can around the class and ask the kids to describe the shape, color, contents, and anything else that they can think of that would quantify the can for a sympathetic scientist on Mars.

2. Point to the hot plate and ask them what this instrument helps us do (it heats objects). Put a cup of water into the can and place it on the hot plate.

3. When the water begins to boil, remove the can from the hot plate (use hot pads), set it on the table, and screw the cap on it tightly.

4. As the can cools it will begin to collapse. You can speed up the process if you hold the can under cold, running tap water.

5. Once the can has collapsed, ask the kids to suggest ways to return the can to its original shape. Eventually, they will decide that you should put the can back on the hot plate. As the contents of the can are heated (keep the cap on), it will once again expand to almost its original shape.

6. Take the can off the hot plate and it will collapse again.

Wrinkled Can

Howcome, huh?

If you were able to magnify a section of air, you would see tiny air molecules bouncing around, bumping into each other and everything else they contact. This "bumping" is known as air pressure. What follows is a demonstration you can do with your class to explain how air pressure works. Place the beach ball on the floor and have four volunteers stand next to it, two on either side. Ask what the kids think will happen to the ball if they push with equal pressure on each side of the ball. Have them try it. They'll notice that as long as they push in the middle of the ball, it might squish a bit but it won't go anywhere. Then ask what would happen if you added two more kids to one side and all kids pushed with the same force. Grab two more kids and demonstrate. The ball should move towards the side with fewer kids.

Explain that the kids pushing on the ball are like air molecules pushing on things. This is called air pressure. High air pressure pushes things towards low air pressure. The side of the ball with four kids represents an area of higher air pressure than the side with two kids. Explain that one way to increase air pressure is to increase the number of molecules doing the pushing.

Get the ball and four kids, just as you did for the last demo. This time remove a kid from one side, ask what will happen, and then have the kids demonstrate. Use this to demonstrate the fact that decreasing the number of molecules on one side is a way of showing the air pressure has been reduced. Reinforce the following: High air pressure pushes things towards low pressure; increasing the number of molecules increases the air pressure at that location. In the same respect, decreasing the number of molecules in a location decreases the air pressure.

can

water

pie tin

Wrinkled Can

On to the wrinkled can. As the air molecules inside the can heat up, they move faster and faster and tend to push a bunch of air molecules out the opening of the can. This results in fewer air molecules inside. As long as the can is on the hot plate, there might be fewer air molecules inside but they're pushing harder than normal. Once you remove the can from the hot plate, you have relatively few molecules inside the can pushing against a whole bunch of air molecules outside the can. In other words, the air pressure outside the can is greater than the air pressure inside the can. The air molecules outside the can win (they have the higher pressure), and the can collapses.

Extensions

If you don't have a plethora of empty unused gas cans around, use 16-ounce or 12-ounce aluminum cans. Beer or ice tea cans tend to work better because they are thinner for some reason than the pop cans, but this may cause some concern in the principal's office. You call this one. Add a couple of teaspoons of water to the can and heat it on the hot plate. When the water is boiling, remove the can with the hot pad and invert it, placing the open side down in a pie tin full of ice water. Same cool crinkling sound. Be sure to recycle the can.

Wrinkled Can

Name_____

Draw a before and after picture of the can.

In the space below, describe why the can collapsed.

_____

_____

_____

_____

_____

_____

_____

_____

Soda Fountain

Blow into a straw on a two-liter bottle and create your own geyser! Ever heard of Old Faithful?

Materials

1 bottle, two-liter
plastic straw
stick of clay
water

Set Up

1. Fill the bottle with water, leaving about two inches at the top.

2. Wrap the clay around the straw one third of the way from the bottom. The clay ball should be just big enough to form a seal in the top of the bottle.

3. Insert the clay ball into the bottle, leaving the tallest part of the straw sticking up. Make sure that the clay forms a good seal in the top of the bottle and that the bottom of the straw is submerged.

The Zing!

1. Place the bottle in front of the students. Have them describe the liquid; also ask if there is anything causing the liquid to move.

2. Instruct a student volunteer to take a deep breath and blow into the straw until he or she can no longer force any air into the bottle. When the bottle starts to win and the student can't get any more air into it, have him or her move quickly to the side. A stream of water should shoot up into the air (and if the student isn't fast enough, all over his or her face).

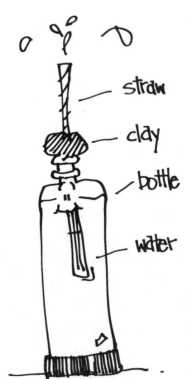

40 ©1989 Rev 1999 The Wild Goose Company WG-3005

Soda Fountain

Howcome, huh?

A quick review of how air pressure works: More air molecules means higher air pressure. Fewer air molecules means lower air pressure. High air pressure pushes things toward low air pressure. By blowing into the straw, you are forcing more air molecules into the bottle. More air molecules means a higher air pressure (compared to the outside air pressure) inside the bottle. This higher air pressure pushes the water in the bottle up and out through the straw.

Extensions

1. You can vary this experiment several ways, including but not limited to:

   • Use different levels of water in the bottle.

   • Change the diameter of the straw used.

   • Try using bottles with different volumes.

   • Change the fluid used.

   • Change the temperature of the water.

   • The upshot of all of these inquiries should be to construct the ultimate soda fountain — one that shoots the most water the highest into the air. Experimenting with these variables one at a time, the kids should be able to produce a couple of real corkers.

2. Have the kids research geysers to find out what makes them erupt.

Soda Fountain

Name_____

1. When someone blows into the bottle, is the air pressure inside the bottle increased or decreased? Why?

   _____

   _____

2. Draw a picture of what happens when the air pressure from the top is released.

Inside Out Bag

Flatten a bag inside a jar or glass and secure it with a rubber band. Try pulling it out of the jar. Easy, you say? Not so.

Materials
    1 jar
    1 plastic bag, 1-quart or smaller

Set Up
No prep. This puppy is a piece of cake.

The Zing!

1. Open up a plastic bag and stick it into a jar. Smoosh the bag against the sides of the jar, taking up as much room as possible. Make sure there is as little air as possible between the bag and the jar.

2. Fold the plastic bag outside over the lip of the jar and secure it with a rubber band.

3. Now, grab the center of the plastic bag and lift. The bag will not come out of the jar, at least not without a lot of effort.

Howcome, huh?

Assuming you fastened the rubber band nice and tight, it will be difficult, if not impossible, for air to travel from outside the jar to the space between the plastic and the jar. As you pull on the bag, you create more space. This means there are fewer air molecules per unit of space in the area between the bag and the jar. There is a conflict between the air molecules outside the jar pushing in on the bag, and the air molecules between the bag and jar pushing out. The more you pull the bag out, the less push there is from the air between the bag and the jar. At some point, even your strongest pull can't overcome the difference in air pressure between the two sides of the bag. Your bag is stuck.

Because the rubber band doesn't create an air-tight seal, air molecules eventually travel from outside of the jar to the space between the bag and the jar, making it possible to remove the bag. Listen for great slurping sounds as the air rushes from one place to the other. Use a wet bag and a wet glass for even more interesting sound effects.

as the bag is pulled up, a low pressure area is created between the bag and jar

Inside Out Bag

Extensions

1. See if the size of the jar makes any difference and once you have tested that, try different shapes to see if that alters the activity.

2. Take an empty two-liter bottle and suck as much air out of it as you can. While keeping your mouth over the opening, stick your tongue inside the bottle and then let go. The bottle will hang from your tongue, and you'll be regarded as quite the cool but weird teacher. There is higher air pressure outside the bottle than inside the bottle. This difference in air pressure pushes your tongue into the bottle, creating a seal that keeps it hanging from your tongue.

Inside Out Bag

Name_____

You are a plastic bag trapped in a jar. Explain five ways you could get yourself out.

1. _____

   _____

   _____

2. _____

   _____

   _____

3. _____

   _____

4. _____

   _____

   _____

5. _____

   _____

   _____

Expando Balloon

Hold a test tube topped with a balloon over a candle flame. The balloon inflates.

Materials
1 balloon
1 candle
1 test tube
1 test tube holder
 matches
 water

Set Up
Get the materials and you're ready to go.

The Zing!
1. Put a couple drops of water into the bottom of the test tube.

2. Stretch a rubber balloon over the top of the test tube, and insert it into a test tube holder. The balloon seals the test tube so that nothing enters or leaves the tube.

3. Light the candle and hold the test tube over the flame until the balloon inflates.

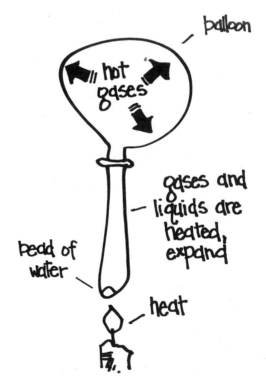

balloon

hot gases

gases and liquids are heated, expand

bead of water

heat

Howcome, huh?
If you've done some of the previous activities, you know that you can increase or decrease air pressure by changing the number of air molecules in a given space. A second way to increase or decrease air pressure is to change how fast the molecules are moving. Suppose you have a bunch of air molecules in a box. They're bouncing around and exerting a certain pressure on the walls of the box. If you can make those molecules move faster, they'll hit the sides of the box harder. Voilà! An increase in pressure. By the same token, you can decrease the air pressure if you can make the molecules inside the box move slower. They won't hit the walls as hard.

In this zinger, you heat the air and water vapor molecules inside the test tube, causing them to move faster and faster. The faster the molecules move, the harder they hit their surroundings. Hit a balloon harder from the inside and it inflates. Higher pressure (inside the tube and the balloon) pushes things (the balloon) towards lower pressure (the air outside the balloon).

Expando Balloon

Extensions

1. Fill one-third of a test tube with baking soda and pour about one ounce of vinegar into a balloon. Without spilling the vinegar into the test tube, slide the neck of the balloon over the opening of the test tube. Tip the balloon up, dumping the vinegar into the test tube all at once. Hold the neck of the balloon as it expands. The increase in air pressure is due to an increase in the number of gas molecules inside the balloon. Baking soda and vinegar combine to make carbon dioxide gas.

2. Put a few small chunks of dry ice into a test tube. CAUTION: HANDLE THE DRY ICE WITH GLOVES — DIRECT CONTACT WITH SKIN CAN CAUSE A COLD BURN. Place a balloon over the opening of the test tube and watch it expand. Dry ice is frozen carbon dioxide. At room temperature, it rapidly transforms into gaseous carbon dioxide. The increase in the number of gas molecules increases the pressure inside the tube and blows up the balloon.

Expando Balloon

Name_____

Put the following events in the correct order by numbering them:

_____ The balloon inflates.

_____ Add a drop of water to the test tube.

_____ Hold test tube over the lit candle.

_____ Place the balloon over the test tube.

_____ The air inside the balloon pushes harder and inflates the balloon.

 ©1989 Rev 1999 The Wild Goose Company   WG-3005

Dancing Dime

Place a dime on a bottle that has just been removed from the freezer. The dime dances the Rumba.

Materials
   1 bottle, two-liter
   1 dime
      water

Set Up
Put the bottle in the freezer for at least a half-hour before doing this zinger.

The Zing!
1. Pull the bottle out of the freezer.

2. Wet your finger with the water and rub it on the surface of the dime. Place the dime on the top of the bottle; the water acts as a seal.

   cold air
   in bottle
   expands
   as it warms

3. Wait for a minute and the dime will begin dancing up and down on top of the bottle.

Howcome, huh?
As for the basics, cold air molecules move slower than hot air molecules. When you place a bottle in the freezer, it's full of slow-moving air molecules. If you bring the bottle out of the freezer and place a dime on top, the air pressure inside the bottle is about the same as the air pressure outside the bottle. However, as the air in the bottle heats up, the molecules inside move faster and push harder on the dime. The dime jumps, allowing some of the air inside the bottle to escape. Fewer air molecules inside the bottle lowers the air pressure, so the dime falls back down. As the air inside the bottle heats up some more, it once again pushes the dime up. This again lets more air out and the dime settles back down. This goes on until the air inside the bottle reaches close to room temperature. If you use a little imagination, the dime looks like it's dancing.

Extensions
1. Accelerate the process by heating the bottle in a tub of hot water or in your breath after a few jalapeños.

Dancing Dime

Name_____

1. Was the air in the bottle hot or cold when the dime started moving?

   _____

   _____

2. As the air in the bottle began to heat up, what happened to the air molecules inside?

   _____

   _____

3. Why does the dime dance?

   _____

   _____

Herculean Index Card

Place an index card over a glass filled with water. Tip the glass and card upside down and stay dry!

Materials

    1 glass
    1 index card
       water

Set Up

Make sure the index card completely covers the opening in the glass.

The Zing!

1. Fill the glass all the way to the brim with water. You don't want it to be spilling over, but more is better with this experiment.

2. Place an index card over the top of the glass.

3. Use one hand to hold the card in place as you flip the container over with the other hand. Remove the hand holding the card so that there is nothing visibly holding it, and everything miraculously stays in place. Oooh Ahhh.

Howcome, huh?

Air pressure is at work again. When you turn the glass and card over, the water level in the glass falls a little. This creates an area of low air pressure at the top (formerly the bottom) of the glass, because you now have the same number of air molecules in a larger space. The air on the outside of the glass, which is pushing up on the water, exerts a large enough force to overcome not just the air inside the glass pushing down, but also the weight of the water. Check out the illustration.

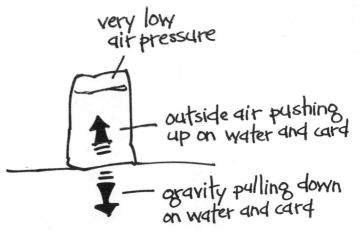

very low air pressure

outside air pushing up on water and card

gravity pulling down on water and card

Extensions

An interesting variation on this experiment is to use an empty soup can. Cut one end out of the can and leave the other end intact, but put a small hole near the top. Demonstrate the experiment for the students and then ask for volunteers. Ask them to repeat what you have done. Unless they cover the hole, the water will fall out of the can. The small hole allows air to enter the can, equalizing the forces. Sneaky but fun.

Herculean Index Card

Name_____

Write down what you thought was going to happen when your teacher turned the card over and describe what actually happened and why.

_____

_____

_____

_____

_____

_____

_____

_____

The Stubborn Balloon

Push a large water balloon into a jar and pull it out again. Think you can do it?

Materials
1 paper towel
1 straw
1 water balloon (homemade)
1 wide-mouthed jar
   matches

Set Up
1. Make the water balloon before the students show up. Add enough water so the balloon just rests on the mouth of the jar but doesn't fall in. If you make it too big or too small the zinger might not work. Once you have the right size, make a back-up balloon just in case you lose the first one in the line of duty.

2. Read through the next section and practice.

The Zing!
1. Place the water balloon on the opening of the jar and ask a student volunteer to push the balloon into the jar. Squeezing the sides of the balloon is not allowed. They will be able to squish the balloon in a little ways, but it will always pop back out. Tell your volunteer not to push too hard or the balloon will explode.

2. Remove the balloon, wad up a sheet of paper, light it on fire, and drop it into the jar. When you are sure that it's burning, place the water balloon back on the opening of the jar. The balloon will bounce up and down a couple of times and then disappear (gulp!) into the jar.

3. After the appropriate applause, hand the jar to your volunteer and ask him or her to pull the balloon out for you. Not possible, is it?

4. After a sufficient number of tries hand the kid a straw. Ask him or her to hold the straw inside the jar next to the edge and, using his or her other hand, pull the balloon out (see diagram).   More applause.

The Stubborn Balloon

Howcome, huh?

This is another zinger that demonstrates the effect of air pressure. The explanations for each part of the zinger are listed below.

The balloon wouldn't go into the jar the first time because the air in the jar was pushing back up on the balloon. The air compresses slightly but not enough to allow the balloon to enter the jar.

By lighting a fire in the jar, you are heating up the air molecules. This increases the air pressure inside the jar. When you place the water balloon on top, the high air pressure inside the jar causes the balloon to jump up, allowing some of the air molecules to escape. This tends to lower the air pressure inside. Also, some of the oxygen molecules in the jar combine with the paper as it burns. This further lowers the air pressure. For awhile, the increase in air pressure caused by heated air balances the decrease in air pressure caused by escaping molecules. Soon, however, there isn't enough oxygen for the fire to burn. As soon as the fire goes out, you're left with only low air pressure inside the jar. Then the outside air pushes the balloon in.

When you try to pull the balloon out, it only goes so far. At that point, the balloon seals the jar opening and prevents any air from entering or leaving. If you pull the balloon out a little farther, you create a low pressure area inside the jar, giving the molecules more room to move around. Fewer molecules per volume means lower air pressure. It's the same situation you created with the **Herculean Index Card** and the **Cheesecloth Cap.** When you insert a straw beside the balloon, you allow air molecules to move freely between the inside and the outside of the jar. With the air pressure equalized, it's easy to pull the balloon out of the jar.

Extensions

This same experiment can be done using a hard-boiled egg and a jar with a smaller opening. You get the egg into the jar the same way, but to get the egg out you must invert the jar, forcing the egg to the opening. Then blow quickly and as hard as you can into the jar. The egg will probably pop into your mouth.

The Stubborn Balloon

Name_____

Write a short story pretending you're the balloon sitting on top of the jar. Describe what you feel and how you wind up in the jar. Also, tell how you are going to escape.

_____

_____

_____

_____

_____

_____

_____

_____

_____

_____

_____

_____

_____

_____

_____

Water Vacuum

Hide a candle under an upside down beaker surrounded with water. Light the candle and watch the beaker fill with water. Ooooooooh. Ahhhhhhhh.

Materials
    1 candle
    1 pie tin, 9"
        food coloring (optional)
        glass, tall and clear
        matches
        water

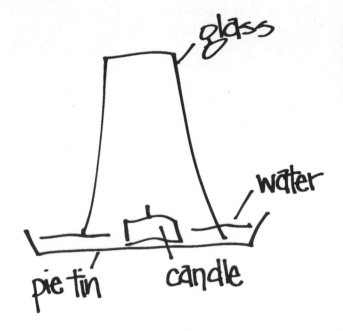

Set Up
1. Add water to the pie tin until it covers the bottom of the tin with a quarter-inch of the wet stuff.

2. Add a couple drops of food coloring. The only reason for doing this is cosmetic; the students can see the results of the experiment better.

3. Place the candle in the middle of the pie tin and put the glass over it.

The Zing!
1. Ask the students to guess what's inside the glass covering the candle (air or oxygen would be appropriate answers). Then ask them what ingredients are necessary for the candle to burn (oxygen, fuel, and heat).

2. Strike a match, lift the glass, light the candle, and place the glass back over the candle. As the candle burns out, the water in the pan will be pushed up into the glass.

3. Repeat the demonstration, having the kids pay special attention to the exact moment the water goes up into the glass. It happens the instant the candle flame goes out.

Water Vacuum

Howcome, huh?

When you place the glass over the lit candle, two things are happening. First, the candle heats the air in the glass, causing the air molecules to move faster. This tends to increase the air pressure inside the glass. Second, the flame uses oxygen (one of the kinds of molecules that make up air) to burn. Decreasing the oxygen molecules tends to reduce the air pressure inside the glass. These two effects pretty much cancel each other out, so nothing happens as long as the candle is burning.

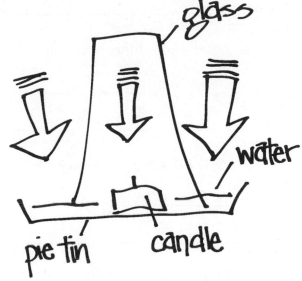

When the candle goes out, you no longer have the source of heat increasing the air pressure inside the glass. You're left with low pressure inside because the flame has removed oxygen molecules. The higher pressure outside then pushes the water up towards the lower pressure inside the glass. Of course, the kids might say that the water was sucked into the glass, but you know that science never sucks. The water gets pushed into the glass by a difference in air pressure.

Extensions

1. Vary the amount of water in the container and see whether or not it affects the outcome of the activity.

2. See whether or not the size of the container affects the outcome of the experiment as well.

3. If you would really like a challenge try the following: Once you have done Extensions 1 and 2, record data and determine whether your results have a relationship between the size of the container and the amount of water in the pan that produces the optimum effect. In other words, how much water with which size container produces the most dramatic effect?

4. About the time you get a handle on that one, add a third variable by altering the size of the candle. Let the creative juices flow.

5. Another way to have a lot of fun is to introduce pure oxygen into the glass. Air is about 78% nitrogen with oxygen comprising about 21% of the remainder. Go to a local hardware store and get a small bottle of oxygen. Fill the glass with water in a sink, insert the nozzle from the oxygen bottle, and fill the glass by displacing water. Cover the bottom of the glass with your hand and then quickly place it over the candle. The flame will burn like crazy and you'll see quite a different response to the change in pressure.

Water Vacuum

Name_____

1. Name the three things a candle needs in order to burn.

   _____

   _____

2. Why was the food coloring added to the water?

   _____

   _____

3. What happened when the glass was placed over the burning candle?

   _____

   _____

4. Why did the water get pushed into the glass?

   _____

   _____

Cloud in a Bottle

Make a cloud appear and disappear with warm water, smoke, and a two-liter bottle.

Materials

1 bottle with cap, two-liter   matches
1 stick of clay   water, warm
1 straw

Set Up

Place the empty two-liter bottle in the fridge for at least an hour.

The Zing!

1. Put about 50 ml (around half a cup) of warm water into the cold, empty two-liter bottle.

2. Light a match and allow it to burn for a second. Then drop it into the bottle. You should end up with a bunch of smoke.

3. Immediately cap the bottle and watch what happens. You should see clouds forming inside.

4. Have a student squeeze the sides of the bottle. This should make the clouds disappear. When the student releases it, the clouds should come back. This can be repeated many times.

Howcome, huh?

The warm water at the bottom of the bottle evaporates into the air inside the bottle. That's a normal thing for water to do, because some of the molecules have enough energy to break away from the rest of the water. When the water molecules are in the air, they form a gas known as water vapor. If this water vapor gets cool enough, it will condense back into water molecules, which is exactly what it does when it encounters the cool air in the bottle. (Remember — you just took the bottle out of the fridge.) The smoke particles help this process along, because they provide a place for the water molecules to collect. The water molecules collecting together on the smoke particles aren't as big and heavy as drops of water, so they just stay where they are. Ladies and gentlemen... clouds.

Cloud in a Bottle

Now, squeeze the sides of the bottle. This puts all the air molecules into a smaller space, increasing the air pressure and also the temperature. This causes the water in the clouds to turn to water vapor and disappear. Pretty much the way it happens in real life. Well, if you use a little imagination!

Extensions

1. Try using cold water instead of warm water in the bottle. Does that increase or decrease the degree of cloud formation? Why?

2. Use a room-temperature bottle instead of one kept in the fridge. How does that affect the cloud formation?

3. Try this demo without the smoke.

4. Mess around with different amounts of water and different temperatures of both bottle and water to see if you can make it rain.

Cloud in a Bottle

Zinger #17

Name_____

1. List the materials you need to make a cloud in a bottle.

    a. _____

    b. _____

    c. _____

    d. _____

    e. _____

2. Tell how to make the cloud in the bottle.

_____

_____

_____

_____

_____

_____

_____

_____

_____

_____

_____

Climbing Test Tube

Insert a thin test tube into a wider test tube filled with water. Turn it upside down and the thin test tube rises into the outside tube.

Materials

1 test tube, thin (16 mm diameter)
1 test tube, fat (20 mm diameter)
1 set of hands
  water

Set Up

Nothing, other than practicing a bunch of times.

The Zing!

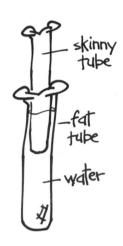

1. Fill the fat test tube half-full with water. Insert the thin test tube (closed side down) into the fat one. It should just rest there on a cushion of water.

2. Use the index finger of the hand holding the fat tube to also hold onto the thin tube, as shown below in the second drawing to the right.

3. Over a sink or outside (or over the room floor if you've practiced and are confident), quickly invert both test tubes so they're again vertical. Let go of the inner tube and watch what happens. See the third drawing. The inner tube should climb up into the outer tube. You can even pull the inner tube down a bit, let go, and watch it climb back up. Gasp! If you're having trouble getting it to work, you're not inverting the tubes fast enough or holding them vertically after the flip.

Howcome, huh?

Before you invert the test tubes, how much air is sitting at the bottom of the fat tube? If you answered "a little," give yourself a star. When you invert the tubes, you'll notice that a little water drips out because gravity is pulling it that way. That leaves a small space at the top (formerly the bottom) of the fat tube. How much air is in that space? Correct answer: "not much." When the water starts flowing out the fat tube, there are very few air molecules in that little space. You pretty much have a vacuum. (In other words, when the trick doesn't work, you'll notice that air bubbles have seeped up into the space between the tubes.)

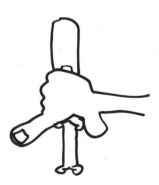

Climbing Test Tube

Now let's look at what's happening to the thin test tube. Gravity is pulling it down. The small air pressure in that space above it is also pushing down, but the atmospheric air pressure from below is pushing up. Because there are lots more air molecules per volume pushing up from below than air pushing down from above, there's an overall push upwards. Hence, a climbing test tube.

Extensions

1. Experiment with different kinds of liquids or amounts of water inside the fat tube.

2. You can observe a similar effect with a syringe (minus the needle). Simply place a finger over the small opening, pull back the plunger, and let go. The difference in air pressure will push the plunger back into the syringe. This works better if the syringe has a little water in it.

3. Speaking of plungers, have the kids figure out how a plumber's helper (toilet or sink plunger) works. It has something to do with creating alternately high and low air pressures.

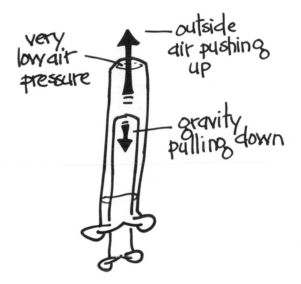

very low air pressure

outside air pushing up

gravity pulling down

Climbing Test Tube

Name_____

1. Explain the best technique for getting the thin test tube to climb up into the fat test tube.

_____

_____

_____

_____

2. Why does the thin tube climb up?_____

_____

_____

3. If you flip the tubes over slowly instead of quickly, the trick doesn't work. Why not?

_____

_____

_____

_____

Bernoulli's Balloons

Suspend two balloons from strings. When you blow between them, they move together rather than moving apart.

Materials
   2 balloons
   2 lengths of string, 18" each

Set Up
1. Inflate and tie off both of the balloons. Attach one string to each balloon.

The Zing!
1. Hold the balloons up for the students to see. You should have one string in each hand and the balloons should be suspended in front of your face. Ask the students to predict what will happen when you blow between the balloons. Will the balloons (a) move away from each other, (b) stay where they are, or (c) move toward each other?

2. Blow between the balloons, and to everyone's surprise they will come together and bump into each other. Hmmmmmm.

Howcome, huh?
This is a demonstration of **Bernoulli's Principle**. The faster a fluid (air is a fluid) moves over a surface, the less pressure it exerts on that surface. If you could see an air molecule hanging out in a still room, it would be bopping around in any direction it wants. This is called **random movement** in physics circles. This random motion of air molecules, which results in them bumping into things, is what causes air pressure. If you take this air molecule and herd it along in a particular direction with a bunch of other air molecules (wind is a good example of this) you are no longer giving that molecule the option of deciding where it wants to go. It now has an agenda. By herding the molecules in a particular direction, you have reduced the amount of push they exert to the side (check out the illustration for further enlightenment).

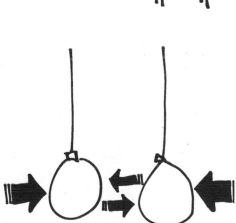

weaker forces in middle allow the balloons to be pushed together.

When the balloons are just hanging there, all of the forces are balanced. When you blow between the balloons, the air rushing into that space pushes very little to the side. The still air on the other side of the balloon pushes harder and this forces the balloons toward each other. This is one of the principles that allows airplanes to fly.

Bernoulli's Balloons

Name_____

1. When air is blown between the balloons, is it moving faster or slower than the air on the outside of the balloon?

_____

_____

2. Does the air pressure on the outer sides of the balloons change at all?

_____

_____

3 Does the air pressure on the inner sides of the balloons change?

_____

_____

4 Why do you think the balloons move toward each other instead of away from each other?

_____

_____

_____

Bernoulli's Funnel

Stuff a funnel with a ping-pong ball and flip it upside down. Blow through the funnel and the ball stays suspended in mid-air.

Materials
1 ping-pong ball
1 funnel
   lips and lungs

Set Up
No set-up necessary. Check for leftovers in the teachers' lounge.

The Zing!
1. Place the ping-pong ball in the funnel, lean back, and try to blow the ball out of the funnel. (Check out the drawing.) Ain't gonna happen!

2. Place the ping-pong ball inside the funnel and hold it there as you flip the funnel upside down. Remove your hand and the ball will fall out. Ask the students for suggestions as to how you can keep the ball inside without using your hand.

3. Once again, place the ball inside the funnel and using your hand to keep it in place, invert the funnel. Place your lips over the small end of the funnel and blow air like crazy into it. After a second, remove your hand. As long as you can blow through the funnel, the ball will stay inside the large part.

Howcome, huh?
This is yet another experiment that demonstrates Bernoulli's Principle. When the air is rushing over the top part of the ball, an area of lower air pressure is created. Since the air pressure on the bottom of the ball is still the same, it pushes up on the ball and prevents it from falling out onto the floor. The diagram on the next page will help wiggle this one into your neural recesses.

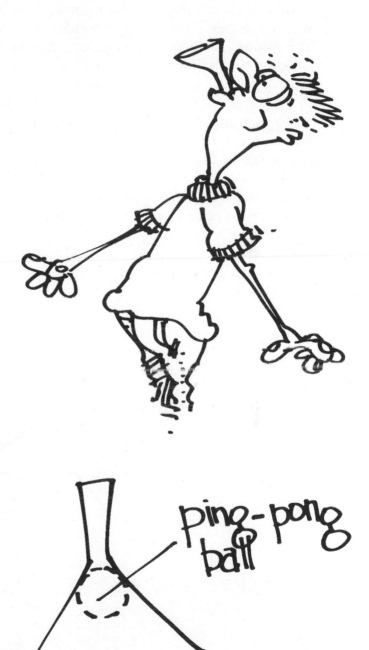

ping-pong ball

funnel

Bernoulli's Funnel

Extensions

1. The same effect can be created by drawing air up into your mouth through the funnel. This may very well have been one of your students' suggestions. The important thing to remember is that you are not sucking the ball up into the funnel. Nothing in science sucks. Remember, you are creating an area of low pressure on the top of the ball by causing the air to rush into the funnel and around the ball. The higher pressure on the bottom keeps the ball from falling out of the funnel.

2. Take a sheet of paper and tape it around the nozzle of a hair dryer to form a cone. Turn the hair dryer on and place the ping-pong ball in the cone. You can now tip the hair dryer back and forth and the ball will not fall out. Keep the angle of the dryer under 45 degrees.

funnel

ping-pong ball

Bernoulli's Funnel

Name_____

1. When the funnel was flipped, did the ping-pong ball fall out of the funnel or stay inside?

   _____

2. What caused the ball to stay inside the funnel?

   _____

   _____

3. Think up three other ways to keep the ball inside the funnel.

   a. _____

   b. _____

   c. _____

Bernoulli's Spool

Using a straight pin, place a paper disk inside an empty sewing spool. Blow through the top of the spool and the disk stays in place. How come?

Materials

- 1 empty sewing spool
- 1 pair of scissors
- 1 piece of paper
- 1 straight pin
- air

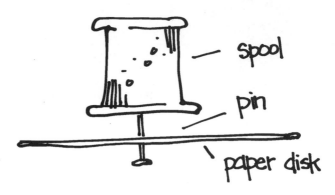

Set Up

1. Cut a paper circle twice the diameter of your spool.

2. Insert the straight pin through the middle of the paper disk.

The Zing!

1. Show the students the spool so they can see that it's hollow inside.

2. Place the disk on the spool with the pin in the hollow part and flip it upside down. Use your index finger to hold the disk in place so that is doesn't fall out.

3. Blow into the end of the spool and release your finger. The paper disk will stay in place instead of falling out onto the floor as everyone expects.

Howcome, huh?

This demonstrates Bernoulli's Principle (see the previous two zingers). When air zips through the spool and out under the disk, it creates an area of lower pressure between the spool and the paper. The air pressure on the outer side of the disk is greater by default and pushes up on the disk. Piece of cake.

Bernoulli's Spool

Extensions

1. Take a sheet of copy paper and fold it in half lengthwise. In some circles this is called the hot dog fold because it is long and skinny like a hot dog bun. Place the sheet on the table so it forms a pup tent. Blow gently through the opening in the middle of the tent. Most people would suggest that the tent sides would expand, but by now you are much too clever to make such a silly, uneducated statement. You already know that the sides actually collapse because the air traveling through the tent is exerting less pressure than the air outside the tent. This difference in air pressure causes the tent sides to be pushed in.

2. More, you say? Take two strips of paper about 1" wide and about 18" long. Old newspaper strips work great. Hold the two strips up about 4 inches apart and blow between them. Some people would suggest that the strips should push apart as you blow through them, but you know better. Remember Bernoulli's Principle? The faster the air moves over the surface of the paper, the less pressure it exerts on that surface. In other words, when you blow through the strips of paper, you are moving the air molecules through to the other side, which decreases the air pressure between them. When the air pressure decreases (less molecules bumping into each other) there is <u>less</u> pressure pushing <u>out</u> on the paper strips. Thus, the strips stick close together, just as the tent sides collapsed inward.

3. OK, one more, but that's it. Grab another sheet of copy paper and place it between two books that are about three inches apart. This should look like a covered carport between two halves of a duplex. Lean down and blow under the sheet, between the books. The paper collapses down rather than blows up. You know why!

Bernoulli's Spool

Name_____

Design three more "Bernoulli inventions" that use the principle of reduced air pressure to hold things in place and move things. Draw and label each of your inventions below and, if possible, try them out!

Soup Can Waterfalls

Punch holes in a soup can, fill with water, and watch out!

Materials

> 1 empty soup can
> tape
> water

Set Up

Begin at the bottom of the can and using a nail, punch a hole one half-inch from the edge. One half-inch above that make another hole, and another half-inch above the second hole, punch a third hole. You now have three holes on top of each other. Should you still be clueless, please consult the drawing above.

The Zing!

1. Cover the holes with a piece of tape and fill the can with water. Ask the kids what they think will happen when you remove the tape. Yeah, sure the water will spurt out, but how? Will all three streams shoot the same distance, or what?

2. Hold the can over a sink or bucket and pull the tape off. Three streams of water will shoot out. The bottom stream should shoot the farthest and the top stream will stay closer to you. Watch out.

Howcome, huh?

Imagine you're a small bit of water inside the can and are ready to head out the top hole. How much pressure is there acting on you? Well, the top of the can is open to the air, so that air pressure is acting on all the water in the can. In addition to that, the weight of all the water above you is pushing on you. Makes you want to head out that hole right away! Now think about the water near the bottom hole. It also has the atmospheric air pressure pushing on it. But since it's lower (than you are) in the can, there's even more water above it pushing down. In other words, the total pressure is higher the lower you go in the can. More pressure means the bottom stream is going to shoot out farther. Makes sense.

If you or any of the kids have been SCUBA (Some Come Up Breathing Air) diving or even just taking a dive to the bottom of a swimming pool, you've experienced first-hand the increase in pressure as you go down into a liquid. The pressure increase is large enough to pop your eardrums if you don't do something to counteract it.

Soup Can Waterfalls

Extensions

You can demonstrate that the weight of the water influences the distance shot out of the can by using another can and punching three holes, all of equal height above the bottom. When the tape is removed, the water will shoot out the same distance. This is because all three holes have the same height of water above them.

Soup Can Waterfalls

Zinger #22

Name_____

Write a poem about the three-holed can demonstration that explains what you saw and what happened. It can be a rhyming poem, free verse, or anything else that you'd like.

_____

_____

_____

_____

_____

_____

_____

Genie in a Bottle

Balance a bottle of cold water on top of a bottle of red hot water. Watch the "red" genie rise into the top bottle. It's scientific magic!

Materials

2 bottles, two-liter   hot water
  cold water   red food coloring

Set Up

Heat some water prior to the zinger.

The Zing!

1 Fill one bottle with cold water and set it aside.

2. Fill the other bottle with hot water, add a couple of drops of food coloring, and mix it up.

3. Invert the bottle of cold water and balance it on top of the bottle of hot water. A red plume of food coloring will begin to wander up into the bottle of cold water.

Howcome, huh?

Heat that is added to a liquid excites the molecules. The molecules, having extra energy to bounce around, begin banging into one another more energetically and expand into the space available. As these molecules expand, they do not increase in number.

If you were to take a gallon of cold water and count all the molecules in it (good luck at keeping your sanity), then take a gallon of very hot water and do the same thing, you would find that the gallon of cold water had many more molecules of water. It is more dense. You could say that a gallon of cold water is heavier than a gallon of warm water.

Since any given volume of cold water is heavier than that same volume of hot water, the cold water displaces the warmer water in the bottom bottle, pushing it up into the upper bottle.

Extensions

Prepare two sets of the bottles prior to class. The first set should be the same as above, and the second set should have dye in the cold water rather than the hot water. Always invert the cold water over the hot. The second set will appear to be opposite the first. The dyed cold water will wander down into the bottle of hot water.

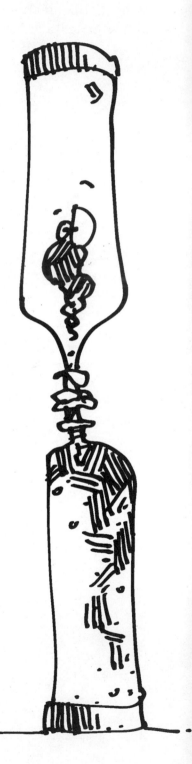

Genie in a Bottle

Name_____

1. When you heat a liquid, what happens to the molecules of the liquid?

_____

_____

2. Is a given volume of warm water lighter or heavier than the colder water?

_____

_____

3. Is the warm water in the bottom bottle or the top bottle at the end of the experiment?

_____

_____

4. Why did this happen?

_____

_____

5. Where did the water in the upper bottle go and why did it go there?

_____

_____

_____

_____

_____

Smoking Chimney

Place a smoking piece of incense next to a bottomless bottle. Light a candle under the bottle and watch the smoke draw up and out through the top.

Materials

    1 candle
    1 bottle, two-liter
      matches
      incense stick or cone
      scissors

Set Up

1. Cut the bottom and top off a two-liter bottle so you end up with a cylinder open on both ends.

2. Cut a small opening (about 2" by 3") in the side of the cylinder at one end. You should end up with something that looks like what that kid to the right is holding.

3. Place the unlit candle inside the cylinder and put the unlit incense cone next to the opening.

The Zing!

1. Show the students the set up.

2. Light the incense on fire and let it burn out until it smolders. Set it near the opening of the bottle and ask the students to observe where the smoke is going. It should be rising straight up into the air. Ask the kids why they think the smoke goes straight up. Look for answers such as "heat rises" or "hot air rises."

3. Light the candle inside the bottle and place the incense near the opening once again. Ask the students to observe where the smoke is going now. The smoke will head into the little trap door and up the inside of the bottle.

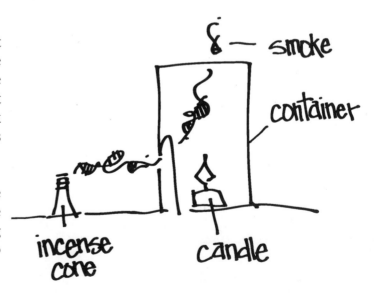

Smoking Chimney

Howcome, huh?

When you heat air, the air molecules move faster. They bump into each other and their surroundings harder than do cool air molecules. If there's room, the more energetic air molecules create more space for themselves. When this happens, the hot air is less dense than cool air. More dense cool air pushing down displaces less dense, hot air pushing upwards. In the smoking chimney, the cooler air outside the cylinder displaces the hot air inside the cylinder in an upward movement, and the smoke from the incense helps you trace its path. Compare this explanation with the one for **Genie in a Bottle**, and you'll see that the two fluids (water and air) behave pretty much the same.

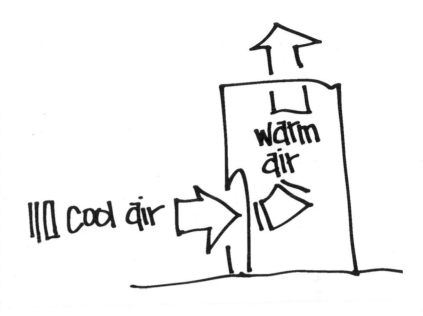

This circulation of air as a result of differing temperatures is known as convection. On a large scale, convection is responsible for many of the weather patterns on Earth. On a small scale, convection helps gliders stay aloft, provides the basis for the design of heating systems, and allows convection ovens to do their thing.

A lot of books and activities will explain convection as hot air rising and cold air rushing in to take its place. Actually, the only reason hot air rises is that it's pushed upward by more dense, cold air. After all, if you had nothing but a big mass of hot air sitting on the Earth, would it just rise up away from the Earth on its own? *Mais, non!* Not unless you reverse the pull of gravity!

Extensions

1. Design an experiment to determine whether the height of the chimney has any effect on the speed of the rising smoke.

2. Figure out a way to detect the movement of hot air up the chimney without using the smoke as a tracer.

3. Is it easier to fly a hot air balloon in cold weather or hot? In the morning or the evening? Why? Prove it! Contact a balloonist and have him or her talk to the class. Did someone say "free balloon rides?"

Smoking Chimney

Name_____

1. Is the air above the candle being heated or cooled?

   _____

   _____

2. Why does the smoke enter the door instead of going straight up into the room?

   _____

   _____

3. Why does the smoke rise?

   _____

   _____

4. Draw a picture of the path the smoke will take as it goes through the convection chimney.

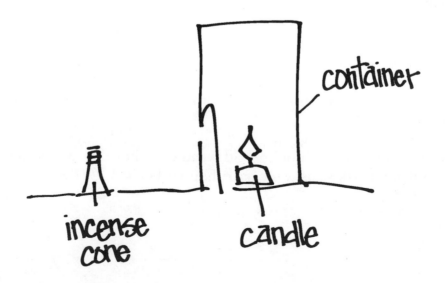

Convection Wheel

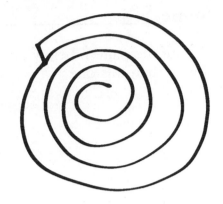

Turn a paper pinwheel using heat generated by a candle.

Materials
    1 candle
    1 pencil
    1 piece of paper
    1 straight pin
       matches

Set Up
You may want to have the class zing along with you instead of just demonstrating. This means you'll have to round up extra materials.

The Zing!
1. Using the pattern from the blackline master on the next page or improvising on your own, cut out a pinwheel.

2. Insert the straight pin into the eraser of a wooden pencil, and balance the center of the pinwheel on the pin.

3. Light the candle and hold the pinwheel 8" above the flame. Move the pinwheel back and forth until you find the current of hot air rising up from the candle. You'll know when you've found it because the pinwheel will begin to spin.

Howcome, huh?
Cooler, more dense air surrounding the candle pushes the warmer, less dense air above the candle in an upward direction. Your friend convection again! Because the spiral hangs down, it always presents an angled surface to the rising air. As the rising air pushes on this angled surface, it causes the spiral to turn. See the diagram. This works on the same principle as pinwheels, windmills, and airplane propellers (note the angled blades on all of these).

Extensions
1. Vary the size of the pinwheel to see how that affects things.

2. Try different materials to see if some will spin faster than others.

3. Alter the cut or modify the design completely to get it to do different things.

Convection Wheel

Name_____

Cut out the pattern below and test it out. Create a design that will spin faster, and one that will spin slower than this one.

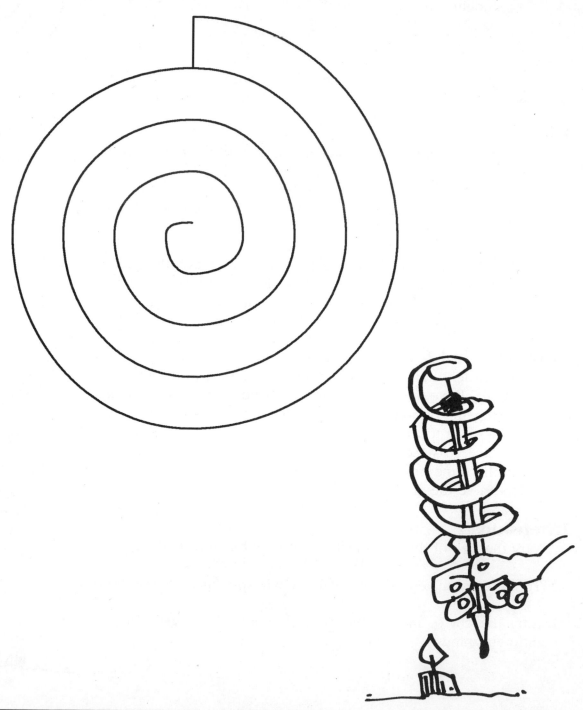

Mechanics

No, not a section on auto repair. In science, mechanics refers to the motion, or lack thereof, of objects under the influence of forces such as gravity and air pressure.

Balancing Nails

Pile six nails on the head of one nail. No glue, no rubber bands, no kidding.

Materials
  7 nails
  1 wooden block (approximately 2" x 2" x 4")

Set Up
Drive one of the nails into the center of the wood block. A ball of clay works fine as a substitute if you can't find wood.

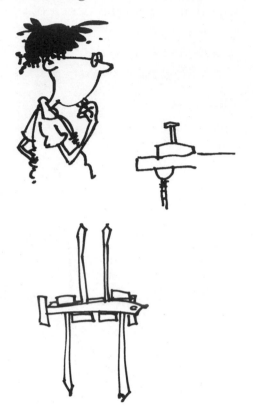

The Zing!
1. Lay out six of the nails in front of the students. Ask the students to suggest some ways the single nail in the block of wood can support all of them. No need to constrain the suggestions and require that they only think of ways that employ balance. You will get glue, rubber bands, and paste in their answers. Direct them to think of ways of balancing the nails without using any binding agents. At this point, maybe it's obvious that having a set of nails and block for each group of two or three kids will turn this into a full-blown science lesson. Only one set-up? Head to step 2.

2. Lay out the nails as shown in the drawing to the right.

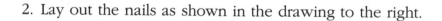

3. Once the nails have been laid out, lift them gently and balance them on the nail, using the picture at lower right for reference. This will astound and amaze your students. You can perform this demo with a variety of different-sized nails. Tool down to the local hardware store and rummage in the bin to make a variety of displays.

Howcome, huh?
To really understand this explanation, you need a visual aid. Grab a regular old carpenter's hammer. Hold it vertically at the end of the handle with the hammer head on top.

Use your other hand to push the head of the hammer a bit to the side. The head keeps on going, falling over until it's below the handle.

center of gravity

Now hold the end of the handle with the head hanging straight down below it. Use your other hand to push the head a bit to the side and it should swing back to its original position. In fact, no matter how hard you push, the head still swings back to the bottom position.

Balancing Nails

The center of gravity of an object is the place where the force of gravity would act on the object if it only pulled in one spot. It's the average position of the weight of the object. In the case of the hammer, the center of gravity is near, if not in, the head of the hammer; the reason being that the head is by far the heaviest part.

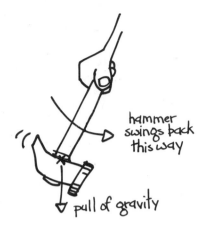

hammer swings back this way

pull of gravity

As a general rule, if you support an object below its center of gravity, it's not very stable. You give it a push and it keeps on falling over. This is what happens when you hold the hammer so the head is directly above the handle. If you support an object above its center of gravity, though, it's very stable. Push it and it goes right back to where it was.

Okay, on to the nails. The first thing that happens when you follow the diagrams is that the four nails hanging down pull the other two together, keeping all the nails from falling apart. The second thing that happens is that the center of gravity of all six nails moves towards the nails hanging down. In fact, when you place the six nails on the one that's in the wood block, the center of gravity of those six nails is actually below the point of support. If you bought the hammer explanation, this means you have a stable configuration. You can bump the nails a bit, and they'll swing around without falling off.

center of gravity is about here, above the cross nails

When you consider the drawing above, you can see that the center of gravity of the nail contraption isn't on or in any of the nails — it's just sort of hanging out there in empty space. If that bothers you, well, that's just the way it is. Maybe you'll feel better thinking about the fact that the center of gravity of a hula hoop is in the center, where there's no hoop at all.

Extensions
1. See how many nails you can get to balance in this fashion. You can get up to 20 nails balanced on the head of a single nail without too much trouble. In fact, more nails hanging down to the side lowers the center of gravity even more, making your contraption even more stable.

2. Field trip it to the local gift shop and see how many executive toys you can find that balance according to the concepts illustrated in this activity. Identify and talk about the center of gravity.

Balancing Nails

Name_____

Draw a picture of the nail configuration you saw your teacher construct. Explain why the nails didn't fall off the single nail like you thought they would.

_____

_____

_____

_____

_____

_____

Spinning Eggs

Spin two eggs and discover which is hard-boiled and which is scrambled.

Materials
    2 fresh eggs
    2 hard-boiled eggs
    1 dish

Set Up
Cook up 2 hard-boiled eggs before class.

The Zing!
1. Place one fresh and one hard-boiled egg on the table and ask the students to tell you which one is which.

2. Spin the eggs. The hard-boiled egg will spin like crazy and the fresh egg will be pretty sluggish about it. Crack the eggs open in the dish to prove your point.

3. Get both of the remaining eggs spinning at the same speed, and then place a finger on each one for a second. The hard-boiled egg will stop immediately. The fresh egg will pause for a moment and then continue to spin.

Howcome, huh?
Inertia is the tendency of an object to keep on doing whatever it's doing, whether it's moving or standing still. A dump truck full of gravel has lots of inertia because it's hard to get it moving, and once it's moving, it's hard to stop. A pea has very little inertia because you can move it with the flick of a finger and stop it without much trouble. In this activity, you are demonstrating rotational inertia. This is basically the same concept, except that it refers to how difficult it is to get an object to rotate or to stop it from rotating.

Visual aid time. Get a relatively light textbook (light in weight, not reading level) and place it flat on a smooth table. Grab a couple of heavy objects no longer than ¼ of the length of the book. The objects can be paperweights, rocks, or wads of bubble gum collected from the bottom of desks. Place the heavy objects as close to the center of the book as you can (on top of each other if possible). Then try to spin the book as shown in the diagram.

Spinning Eggs

Note how difficult this is. Then move the weights to the edge of the book and spin again.

This should be lots harder. What this shows is that objects have more rotational inertia when their weight is distributed towards the edge than when it's near the center. Ice skaters know this because they spin much faster when they bring their arms in close to their bodies as opposed to having their arms outstretched.

Yeah, but you're dealing with eggs, not ice skaters. When you spin a hard-boiled egg, all the innards stay in place. When you spin a fresh egg, the liquid inside moves to the edge, making the egg much harder to spin. When you briefly place your finger on each egg, the liquid in the fresh egg keeps rotating inside, so there's still rotational motion after you let go.

Extension

Find another chair that spins rather easily. Have a student sit in the chair with feet off the floor and a heavy weight in each hand. Instruct the kid to stretch his or her arms out, and start him or her spinning slowly. Then tell the kid to bring his or her arms in quickly. As the kid spins out of control, explain how there's nothing like actual experience to help understand something like rotational inertia.

Spinning Eggs

Name_____

1. What happened when the fresh egg was spun?

   _____

2. What happened when the hard-boiled egg was spun?

   _____

3. What caused the difference ?

   _____

   _____

4. Describe what happens if you place your finger on a hard-boiled egg that is spinning and let go.

   _____

   _____

5. Describe what happens if you place your finger on a fresh egg that is spinning and let it go.

   _____

6. What do you think causes the difference?

   _____

   _____

Balancing Act

Balance a full-sized broom in the palm of your hand.

Materials
1 stick of clay
1 wooden dowel

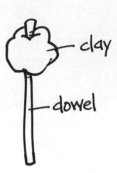

clay

dowel

Set Up
Roll the clay into a ball and skewer it on to one end of the dowel. If you haven't done **Spinning Eggs** yet, do it first. Once you understand that activity, this one's a piece of cake.

The Zing!
1. Show the dowel to the class clay-side down. Challenge one of the kids to balance the dowel in the palm of their hand for as long as possible. Their palms must be held flat.

2. After the kid struggles and fails miserably (poor kid), take the dowel and balance it in your palm. You'll be successful because you'll place the end with the clay away from your palm.

3. Get ready for some more fun! Try balancing a broom or baseball bat. The heavy end of the broom corresponds to the end of the dowel with the clay on it. If you try a baseball bat, the hitting (fat) end of the bat corresponds to the clay end of the dowel.

Howcome, huh?
In **Spinning Eggs**, the farther from the center of rotation the weight of an object is distributed, the more rotational inertia it has. A wheel with all of its weight on the rim is much harder to rotate or stop from rotating than a wheel with its weight in the hub (center). Now apply this to the dowel and clay. Think of the dowel falling over as a rotation, with your palm as the center of that rotation.

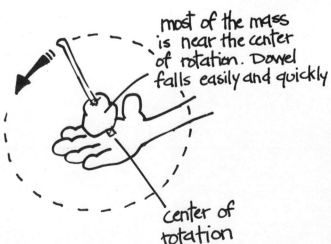

most of the mass is near the center of rotation. Dowel falls easily and quickly

center of rotation

90

Balancing Act

The force of gravity causes the dowel to fall over. With the clay next to your hand, the dowel and clay have a small rotational inertia (most of the weight near the center of rotation) and it's easy for gravity to do its thing. In other words, it's very difficult to balance in this position.

When the end with the clay is away from your hand, the dowel and clay have a much larger rotational inertia. It's difficult to make it fall (rotate). Because of this sluggishness, you have more time to adjust the position of your hand and keep it balanced.

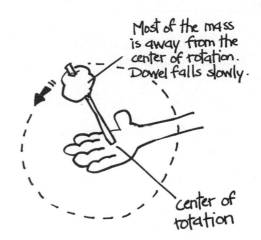

Most of the mass is away from the center of rotation. Dowel falls slowly.

center of rotation

Extensions

1. Do a full-blown experiment to determine the relationship between the position of the clay on the dowel and the time of balance. Try this for at least six different clay positions.

2. Invite a juggler to class to explain how he or she uses physical principles to balance objects. Chances are the juggler won't understand the science, but you can deal with that once you learn some tricks of the trade.

3. Invent a carnival ride that employs changing centers of gravity and changing rotational inertia.

Balancing Act

Name_____

1. How long did it take for the dowel to fall (rotate) when you placed the ball of clay in your hand? (Keep your palm flat.)

   _____

2. How long can you balance the ball of clay when you place the end of the dowel in your hand?

   _____

   _____

3. Move the ball halfway down the dowel and try to balance it. How easy is it? Does the ball fall (rotate) and if it does, record the time it takes for this to happen.

   _____

   _____

4. Make a chart showing the relationship between the position of the clay ball and how long it takes gravity to cause it to rotate.

   _____

   _____

5. How can you (or the juggler) compensate for the force of gravity that keeps things falling (rotating) to the center of the earth?

   _____

   _____

6. How does the design of the human body allow you to stand without effort?

   _____

   _____

Balloon Rocket

Race a balloon rocket from your desk to the ceiling and entertain all!

Materials
- 1 balloon
- 1 straw
- 1 very long piece of string
- hot air
- tape

Set Up
1. Attach a string to the ceiling with tape on the far side of the room.

2. The string should be long enough to stretch all the way to your desk.

The Zing!
1. Feed the string through the straw. Attach 3 pieces of tape to the straw. Use the diagram above as a guide.

2. Blow the balloon up but do not tie it off, and attach it to the straw with the tape.

3. Have the students count to three and release the balloon. It will race up the string. Cheering and wild applause.

Howcome, huh?
The venerable Sir Isaac Newton put forth the idea that for every action there is an equal and opposite reaction. In this case, the gases escaping from the back of the balloon rocket thrust the rocket forward. For one reaction there is another, opposite reaction.

Another way of stating this law (known as Newton's Third Law) is that when object A exerts a force on object B (the action), Object B exerts an equal and opposite force on object A (the reaction). With the balloon, the balloon pushes on the air that escapes out the back (action) and the air pushes back on the balloon (reaction). The result is that the balloon accelerates. (This last sentence is actually Newton's Second Law, but who's counting?)

Extensions
It is fun to add additional balloons (increased thrust) and/or additional weight to see how it affects the speed at which the rocket travels up the string.

Balloon Rocket

Name_____

Design an experiment that tests increased thrust (extra balloons) and increased load (extra weight) on the performance of the balloon rocket. List the materials that you will need and the procedure that you'll follow to test your ideas.

_____

_____

_____

_____

_____

_____

_____

_____

_____

_____

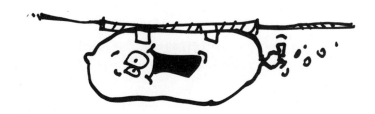

The Straw Mystery

Blow through a flex straw and it zips to the side. Inhale through the same straw and nothing happens. What gives?

Materials
1 flex straw for each kid in the class

Set Up
Try the demo first so you can tell the kids how to do it.

The Zing!
1. Have the kids bend their straws until they form a right angle.

2. Tell the kids to bend at the waist and facing the floor, hold the long end of the straw gently between their lips (just lips, no hands). Check out the drawing.

3. Tell the kids to blow through the straw. The straws should zip to the side. If not, the kids are holding the straw too tight with their lips, or they're holding on to it with their hands.

4. Ask the kids to predict what will happen when they inhale through the straw. They'll most likely say the straw will go the other way. If you wish, inform them that it's impossible to suck through the straw because science never sucks. Really!

5. No matter how hard they inhale, the straw shouldn't budge.

Howcome, huh?
When you blow through the straw, the air goes down to the bend, where the shape of the straw forces the air to the side, out the opening. Because the straw is pushing air to the side, the air pushes back on the straw (Newton's Third Law), shoving it in the opposite direction. It's basically the same principle that makes an air-filled balloon fly around the room when you let go.

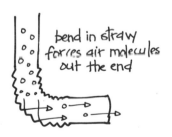

bend in straw forces air molecules out the end

The next drawing shows the path of the air when you inhale. Let's look at what happens. By changing the position of your diaphragm and allowing your lungs to expand, you create a low pressure area inside your mouth. The atmospheric pressure outside the straw then pushes the air (or milk or whatever) in the straw up into your mouth.

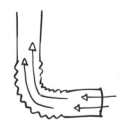

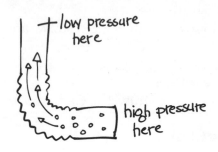

low pressure here

high pressure here

It might seem from the drawing that inhaled air would bang on the bend in the straw and push it sideways. But as soon as the air reaches the bend, atmospheric pressure immediately pushes the air up towards the low pressure area in your mouth. Any push the incoming air gives to the straw gets distributed over the length of the straw rather than right at the bend.

Extensions

Ask the kids whether or not a straw will work on the moon. The quick answer is no, it won't work, because as soon as you take off your space helmet to use the straw, you'll be dead. The longer answer is: There is virtually no atmosphere on the moon, and no atmospheric pressure! You can lower the air pressure inside your mouth all you want, but without outside air pressure to push liquid up into the straw, nothin's gonna happen. As you know, science never sucks.

The Straw Mystery

Name_____

1. What happened when you blew through the straw?

_____

_____

_____

2. What happened when you drew air up into the straw?

_____

_____

_____

3. What caused the difference?

_____

_____

_____

Helicopter

Make a twirling chopper and get out of the way!

Materials

1 pair of scissors
1 piece of paper with pattern on it (see page 100)

Set Up

Have the materials ready for the students, and try making one first.

The Zing!

1. Make enough copies of the student page for the entire class. Have the kids cut out the helicopter pattern along the outer solid black lines; save the rest of the paper for additional designs.

2. Tell the kids to cut along the solid black vertical line, and fold down the two resulting flaps in opposite directions. It should look like a letter T with bunny ears. The bunny ears are the rotors of the helicopter.

3. Have the kids cut the two remaining solid lines. Fold the long sides into the middle as you would a legal letter.

4. Finally, fold the bottom of the helicopter up about one-half inch.

5. When the kids drop their helicopters from an outstretched arm, they (the helicopters, not the kids' arms) should twirl to the floor.

6. Ask the kids whether the copters twirl clockwise or counterclockwise. You'll get conflicting answers, and not just because half the kids don't know what you mean by clockwise. With your question as a catalyst, the kids will eventually figure out that changing the position of the rotors changes the direction the copter spins.

Howcome, huh?

Pretty basic stuff. As the copter falls, the rotors push against the air. The air pushes back, slowing the descent of the copter. The copter spins because the air molecules push on the rotors and cause it to rotate. Check back to the zinger **Crazy Spinner** for an explanation of how forces, in the form of torques, cause rotations.

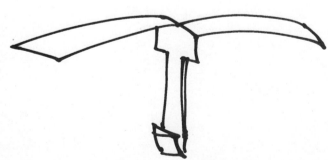

Helicopter

Extensions

Have a contest to see who can make:

a) the biggest helicopter that still flies

b) the smallest

c) the fastest

d) the slowest rotating

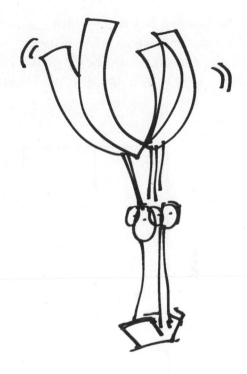

Make a helicopter that has

e) a tubular body

f) a square body

g) an elongated body

h) 3 rotors instead of 2

i) 4 rotors, or 6 rotors, or how many can you add?

j) Is it possible to construct a helicopter with rotors on both ends? Certainly.

By now you should be getting the idea that it is time to open it up to the students and let them run wild with the ideas. You can have all kinds of contests with success for everyone.

Helicopter

Name_____

Using the pattern below, construct a helicopter that will fly. Once you have done that, create three more helicopter designs of your own. Make the rotors (blades) longer or shorter, thicker or thinner. Change the length of the body of the helicopter; test all of your ideas. If you fail, use it as positive feedback. Don't give up. It took Thomas Edison over 6000 tries to invent the light bulb.

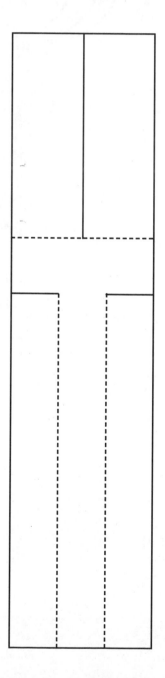

Jet

London in three hours? Create your own ballistic airplane.

Materials

1 piece of white paper for each student.

The Zing!

1. Fold the paper in half lengthwise and unfold it again.

2. Take the upper-right corner of the paper and fold it over to the center of the paper. Do the same thing with the other side of the paper. Your airplane will look like a very tall house right now.

3. Grab the side of the paper (where the gutter of the house would be) and fold it to the center of the paper once again. Do the same thing with the other side. Your airplane should now look like an alpine cabin.

4. Fold the plane into itself so it looks like a giant wedge.

5. Fold the upper edge of the wedge down to the bottom of the wedge; these are your wings. You now have a very streamlined airplane. You may want to add a piece of tape to the top of the plane where the folds come together. It gives your plane more stability.

6. Give it a toss and be sure to watch out for eyeballs.

Howcome, huh?

This paper airplane glides along because as it falls, it pushes on the air and the air pushes back (Newton's Third Law), slowing its descent. With real airplanes, as the air travels over the wing, there is actually less air pressure on top of the wing than there is underneath it. Because there is less pressure, the plane is actually lifted or pushed up. This is another example of **Bernoulli's Principle.** As you already know, there are several activities that explain this principle in this book.

Extensions

Go for the airplane contest; it gives you a wonderful opportunity to get a little sun in the name of academia. Try a couple of different contests so you can spread the accolades around. A suggestion:

a) distance   b) acrobatics   c) design

Jet

Using the patterns below, fold a jet and give it a toss. Once you have mastered this, try other designs of your own.

a

b

c

d

e

102

Acrobatic Plane

Design an airplane that demonstrates the effects of flaps on performance and stability.

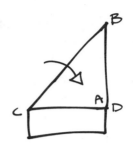

Materials
- 1 pair of scissors
- 1 piece of paper

Set Up
Have enough paper for everybody and make sure that lunch gets eaten! You don't want any extra calories running around the building.

The Zing!

1. Fold point A down to point D. Crease and unfold.

2. Fold point B down to point C. Crease and unfold.

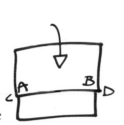

3. Make a hot dog fold (long side to long side). Crease and unfold.

4. Now, fold point A down to point C and point B to point D. Crease and unfold. The fold lines sort of look like a star. You can turn the paper over and refold all the lines if you want.

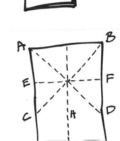

5. This next step is a little tricky but it's very important. Push a little on the center of the "star" to bring points E and F closer. Pinch E and F together and fold them down to point H. Carefully flatten the paper along the fold lines to form a triangle. Point G is now at the top of the triangle.

6. Fold point G down to point H. Notice that the lower corners of the "roof" (C and D) can be lifted. To keep your plane from acting like a helicopter, tape these corners down or tuck them under to point H.

7. Turn the plane over and fold it in half along the center crease. Fold the wings downward on both sides, as shown. Fold each wing tip upward a half-inch so you have stabilizers.

Acrobatic Plane

8. OK, now you're ready to fly! Hold the plane on the underside toward the nose and toss it forward with a flick of your wrist. Outside, your plane will fly long distances in broad loops if there's not too much wind. If the wind is up, throw your plane into the wind. Point it slightly upward. Flying the plane inside can be fun if you have a fairly large space to use. Uh, no, the school library is not a good location.

slits

9. Cut four small slits into the backs of the wings for "elevators."

Howcome, huh?
The flaps at the back of the plane divert the flow of air over the wing, either up or down. Using Newton's Third Law, if the flaps push on the air, the air pushes back, causing the plane to veer right or left.

Extensions
Even without flaps, this plane tends to be more acrobatic than the jet design. That's because it presents a greater surface area to the air molecules and is more susceptible to variations in forces from those molecules. The addition of flaps results in turns, loops, and rolls. As the air molecules hit the flaps, they exert what is known as a **torque** on the plane. To see the difference between a force and a torque, place a pencil on a smooth surface. Use your finger to push the pencil from the side, and in the exact middle. The pencil should slide across the surface without rotating. Now push from the side, but at one end of the pencil. In addition to sliding across the surface, the pencil should rotate. In the first case, you exerted a force on the pencil. In the second, you exerted a torque. So a torque is nothing more than a force applied so the object rotates.

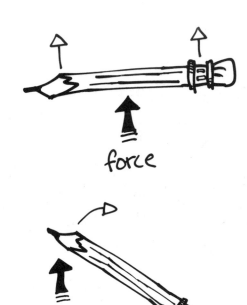

force

torque

The addition of flaps to your plane allow the air molecules to push "off center," causing all sorts of rotations.

Acrobatic Plane

Name_____

Using the pattern below, construct an acrobatic plane and then experiment with bending the tail section up or down. Write a brief summary of what you find out.

_____

_____

_____

_____

_____

_____

_____

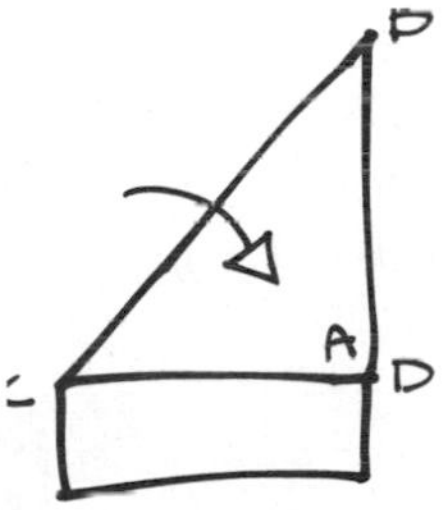

1    2 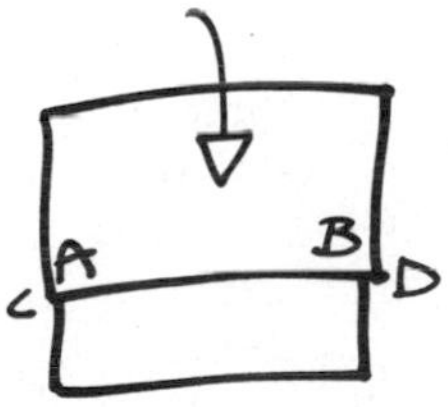   3 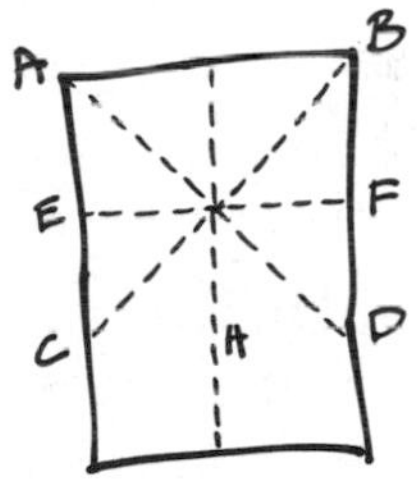   4

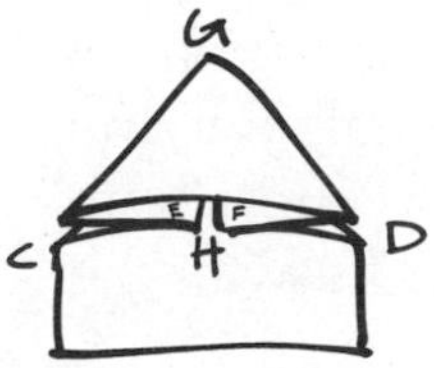

6    7    8 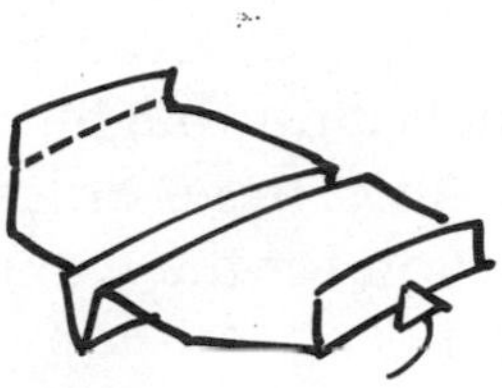   9

Loop Plane

Construct a plane using two loops and a straw. It actually flies!

Materials

1 pair of scissors
1 piece of paper with design to cut out (page 107)
1 straw
1 strip of tape, 6" to 8" long

Set Up

Have the necessary materials ready for all of the students. Don't you hate these extensive preparations?

The Zing!

1. Have the students cut out both of the paper strips on page 107.

2. The tape should be torn into quarters, each piece about 1 to 1.5 inches long.

3 Bend the paper strip into a loop and use the tape to stick the two ends together. You should have two loops, one larger than the other.

4. Place one piece of tape at the very end of the straw. Make sure that the tape is sticking out so the straw and the tape look like a capital T. Place the other strip of tape at the other end so that your straw now looks like a capital I.

5. Slide one loop over the end of the straw and fasten it to the tape there. Do the same thing on the other side of the straw, so now you have this really funny looking plane. Toss that sucker.

Howcome, huh?

The way this plane and all other paper gliders work, is that that they glide on a cushion of air. As the plane starts to fall, it pushes against air molecules which push back to keep the plane aloft. Because one loop is larger than the other, this loop has more surface area and gets pushed harder. When the large loop is in back, the flight is stabilized because the larger push on the back is compensated for by the fact that the front loop funnels air through the large loop so not as much pushes up on the back. It all equals out and the plane flies more-or-less on a level path. When you fly the plane large-end first, there's no compensation for the extra push on the large end, so the plane flips over and begins to fly with the small end first.

Extensions

The applications here are endless. Let the students run wild with ideas of adding loops, changing the size and length. Basically, an "anything goes" policy is best. Have contests to see who can make planes with the most loops; the plane that flies the farthest, highest, longest; and who can make the smallest plane or the biggest plane.

Zero to Einstein in 60™   **106**   ©1989 Rev 1999 The Wild Goose Company WG-3005

Loop Plane

Name_____

Using the pattern below, cut out the loop plane, assemble it, and give it a toss. Once you have experimented with this design, modify it by adding loops, changing their position on the plane, and generally letting your imagination run wild.

Energy Transfer

Much of what makes the world go round has to do with energy transferring from one object to the next or energy changing forms. The first two activities in this section deal with the generation, transfer, and absorption of heat. The second two are unexpected examples of motion energy transferring from one object to the next. Short section, many surprises.

Cool Business Card

Hold a business card in a candle flame and . . . why doesn't it burn?

Materials
1 bucket of water
2 business cards
1 candle
1 old crayon
matches

Set Up
Cut up about one-eighth of a crayon into small pieces or shavings. You choose the color. Have the bucket of water handy just in case you get a larger fire than you intend.

The Zing!
1. Light the candle and ask the class what will happen when you hold a business card over the flame. If someone thinks the card won't burn, keep a close watch on that kid for the rest of the year.

2. Hold a business card just above the flame. It should catch fire within a few seconds.

3. Place your pile of crayon shavings on top of a second card. Ask the kids what will happen when you hold this card over the flame.

4. Hold the card over the flame so the crayon shavings are directly above the flame. In a short time, the shavings will melt. The business card won't even be scorched. This might be a good time to caution the kids against trying this without adult supervision. No sense ruining your day with a burned-down house and a lawsuit to boot.

Howcome, huh?
In order for paper to burn, it has to reach a temperature of 451 degrees Fahrenheit. Crayon melts at a lower temperature than that. In this demo, heat from the candle transfers to the paper, and heat from the paper transfers to the crayon. As the crayon melts, it absorbs heat from the paper just as fast as the paper absorbs heat from the flame. The paper never absorbs enough heat to reach 451 degrees, so it doesn't burn.

Extensions
Experiment with different kinds of materials on the card. Use plenty of caution as you do this.

Cool Business Card

Name_____

1. Make a chart showing the different materials you applied to the cards. Record the time it took for these cards to catch on fire.

_____

_____

2. Which materials prevented the paper from burning?

_____

_____

3. Which materials made the paper burn faster? Why?

_____

_____

_____

_____

Paper Cup Pot

Boil water in a paper muffin cup.

Materials

    1 candle
    1 paper muffin cup
       matches
       banquet fork
       water

Set Up

Have all the materials ready to go, and that's about it.

The Zing!

1. Shishkabob the muffin cup with the fork and put just enough water in the bottom to cover it. It is important to make sure that the entire bottom of the cup is covered, or the cup will catch on fire.

2. Light the candle and hold the paper cup just above the flame. You can speed up the process just a bit if you move the cup back and forth. After a couple of minutes, the water in the cup will begin to boil.

Howcome, huh?

This is similar to the previous zinger. The water absorbs heat from the cup just as fast as the cup absorbs heat from the flame. The reason for this is not that the water is melting, though (no, duh)! Every material has a property known as **specific heat**. For materials with a low specific heat, a small absorption or release of heat gives rise to a large change in temperature. Aluminum has a low specific heat, so it increases or decreases temperature rapidly when it gains or loses even a small amount of heat. That's why aluminum pie tins are safe to touch almost immediately after you take them out of the oven; they cool off rapidly. Water, on the other hand, has a high specific heat. It can absorb or release lots of heat without changing temperature very much. The water in the cup, for example, absorbs lots of heat while keeping the cup at a temperature below 451 degrees Fahrenheit. By the way, water's high specific heat is what makes the climate near oceans so temperate.

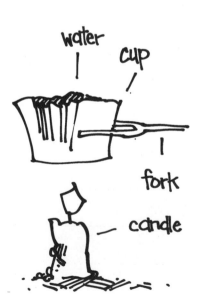

water
cup
fork
candle

Extensions

Adaptations for this experiment include adding different quantities of water to the cup, or using different kinds of liquids (demonstrating the different specific heats).

Paper Cup Pot

Name_____

Write a short story. Pretend that you are a paper cup in a burning house. When the fire fighters break through the door, you are the only object in the house that hasn't burned. Your rescuers are amazed, but you can explain how you escaped disaster.

_____

_____

_____

_____

_____

_____

_____

_____

_____

_____

_____

_____

_____

Wobbling Washers

Bump two washers into action.

Materials

2 lengths of string, 2 feet
2 washers
1 length of string, 3 feet

Set Up

Have everything ready to go, that's it.

The Zing!

1. Tie the 2 two-foot strings to the 3-foot string. They should be evenly spaced along the string and then hang down.

2. Tie a washer to the end of each 2-foot string.

3. Have two of your student volunteers hold each end of the string so it's taut.

4. Pull one of the washers to the side and release it, letting it swing freely. Observe what happens.

Howcome, huh?

When you hang a washer from a string, you have a **pendulum**. Obviously, the energy from the first pendulum transfers to the second, back to the first, and so on. The string holding the washers is the transfer culprit. But there's more to it than that. What you'll find is that the lengths of the two strings hanging down have to be nearly identical. Even small differences in length mess up the transfer of energy. That's because every pendulum has a definite frequency (number of swings in a given time period) depending on the length of the pendulum. Pendulums of like frequency readily transfer energy back and forth, but pendulums of different frequency aren't as effective. The fancy name for what you observe here is **resonance**. It's the same phenomenon that makes it possible for a singer to break a wine glass by hitting just the right note.

Wobbling Washers

Extensions

1. Show that the weight of a pendulum doesn't affect this demo by putting four washers on one string and only one on the other. It should work just fine.

2. Show that the length of the pendulums affect this demo by using strings of different length.

3. Add a third pendulum for more excitement.

Wobbling Washers

Name_____

Design an experiment to decide whether or not the length of the washer string has any effect on this activity. List the materials you will need and how you are going to experiment.

Materials

_____

_____

Procedure

_____

_____

_____

Data

_____

_____

_____

Conclusion

_____

_____

_____

_____

_____

Wilberforce Pendulum

Demonstrate energy transfer with a Slinky™, a dowel, and some clay. Looks like magic!

Materials
    1 dowel (about twice as long as the diameter of the Slinky™)
    1 Slinky™ or Slinky Jr.™
    1 stick of clay

Set Up
If you want to just wow the class with your expertise, read through the procedure and get your pendulum working before you show it to the class. If you want to involve the kids and include a few failures, go ahead and wing it.

The Zing!

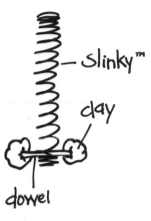

1. Insert the dowel between two sections of the Slinky™, near the end of the spring.

2. Form 2 balls of clay and attach one to each end of the dowel.

3. Hold the Slinky™ at the opposite end and let it hang freely. Start the Slinky™ bouncing up and down and watch what happens. If you're incredibly lucky, the up-and-down motion of the Slinky™ will diminish, and the dowel and clay will start to rotate back and forth.

   The rotational motion will then die down and the up-and-down motion will start up. This transfer will go on for awhile.

4. Assuming you're not lucky, alter the amount of clay on the dowel and the position of the dowel on the Slinky™ until you get the described motion. It's impossible to determine the exact amount of clay or position of the dowel, because both clay and dowel weights vary a lot. You wouldn't want to spoil your fun anyway, right?

Wilberforce Pendulum

Howcome, huh?

This is similar to the previous zinger, so go back and read that explanation if you haven't already. A weight bouncing up and down on a spring has a **resonant (natural) frequency**, just as a pendulum does. Something twisting back and forth while hanging from a string or wire (known as a torsional pendulum) also has a resonant or natural frequency. With your dowel and clay attached to a Slinky™, you have both of these combined. By changing the amount of clay on the dowel and the position of the dowel, you are altering the frequencies of the two motions (up-and-down and rotating). When you hit on just the right combination, the two frequencies match and energy readily transfers from one motion to the other. The energy transfers back and forth because as a Slinky™ moves up and down, it also rotates back and forth. Watch closely and you'll see the rotation.

Extensions

1. Combine this activity with the previous one so you have two Wilberforce pendulums hanging from a string. See what kind of energy transfers you can get.

2. Find out how the Wilberforce pendulum got its name.

Wilberforce Pendulum

Name_____

In the space below, describe the motion of the Wilberforce pendulum.

_____

_____

_____

_____

_____

_____

_____

_____

_____

_____

_____

_____

_____

_____

_____

Sound and Light

This section contains activities that deal with . . . ummmmm . . . sound and light. They're a natural combination because both sound and light travel as waves.

Dueling Speakers

Produce a pattern of loud and soft spots around the room with two speakers that emit the same pitch.

Materials

1 receiver/amplifier plus two speakers or 1 boom box with two speakers

Set Up

Set up your sound system, whatever it may be, so the two speakers are right next to each other, and facing out into the room. They should be as close to the height of the kids' heads as possible, so your desk might be a good place. Punch whatever buttons are necessary to receive an AM radio station. If you have a stereo receiver with a "mono-stereo" switch, set it to mono. Turn on the radio and move up and down the dial until you get one of those really annoying high-pitched whines that kind of go "ooooh-weeeeeeeee-oooooooooh" as you move the dial slightly. Set the dial so you get as loud a whine as possible with as little music or talking as possible. Why yes, it is rather obnoxious. Believe it or not, you're set once you've found the best whine possible.

The Zing!

1. Tell the kids you have something very important for them to listen to. Turn on your sound system and let them feast their ears on the loud whine you found.

2. Have them concentrate on the whine and, as best they can, ignore any other sounds coming from the speakers. Tell them to close one ear and move their heads very slowly from side to side. They should hear the whine get louder, softer, and louder. The sound might actually disappear in some locations. It's a pretty cool effect, so let the kids who hear the difference right away help the others find it.

3. If you can't identify the loud and soft spots, try adjusting the separation of the speakers. If that doesn't work, try finding a higher or lower-pitched whine.

Howcome, huh?

When you hear a sound, it's the result of tiny air molecules moving around and slamming into your eardrum. At the source of the sound (in this case, the speakers) something vibrates, causing the air molecules around it to vibrate. This vibration then spreads around the room as a sound wave. Sound waves spread out much like water waves spread out from a rock dropped in a pool of water. Sound waves aren't exactly like water waves (see the extensions for a more accurate picture), but that model will work here. As you can see from the drawing, waves have both crests (high points) and troughs (low points).

Dueling Speakers

When two waves meet, they combine, cancel each other out, or do something in between. For example, when two crests come together, they add up together to make a higher crest. Two troughs combine to make a lower trough. But when a trough and a crest come together, they tend to cancel each other out.

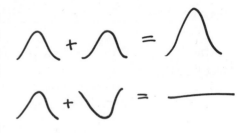

When you have two speakers side-by-side emitting the same tone (that's the whine), the separate sound waves from those two speakers form a pattern. This pattern indicates where in the room they combine and where they cancel each other out. The pattern looks something like the next drawing.

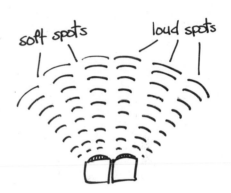

When you move your head from side-to-side, your ear hears places where the waves add to each other and places where the waves cancel each other out. That's why the whining sound appears, disappears, and then appears again.

Extensions

1. Here are two ways to demonstrate what sound waves look like. One tastes better than the other! Get a Slinky™ and stretch it out in front of the class. Rapidly move one end of it in and out.

   When you get the motion right, you'll see waves moving along the length of the Slinky™.

   The tastier demonstration is to cook up a batch of Jell-O™, in a long, column-shaped mold such as a bread box. Remove the Jell-O™ from the mold (once it sets — duh) and place it on a tray. A whack at one end will send a wave traveling to the other end. Waves like this, and sound waves, are called **longitudinal waves**, as opposed to water waves and waves on a string, which are **transverse waves**. For you nitpickers out there, water waves aren't exactly transverse waves because they involve elliptical motion.

2. Invite an engineer or a representative of your local auditorium to talk about how people design auditoriums so as not to have a bunch of "dead spots" where sound waves cancel out.

3. Have the local high school or college physics teacher bring a laser and a set of double slits to your class. When you send light through a double slit, you have a situation analogous to the two speakers. You end up with very cool looking light and dark areas where the light from the two slits adds and cancels out.

Dueling Speakers

Name_____

In the space below, draw a picture of the pattern of loud and soft spots created in your room by the two speakers.

Sound Beats

Pluck two guitar strings tuned at the same pitch and get a cool wah-wah effect.

Materials

1 guitar (electric with amplifier is best) or 1 violin or one of any kind of stringed instrument.

Set Up

Spend a little time to try and scrounge up an electric guitar, because this zinger is much more effective when it's nice and loud. Remember the first rule of rock bands: If you're loud enough, you don't have to be good.

The Zing!

1. Tune two of the strings of the guitar so they're almost, but not quite, the same pitch.

2. Pluck the two strings at the same time and listen for a wah-wah sound, in which the notes get louder, then softer, then louder, etc. If you don't hear that effect, the strings are either too far apart in pitch or you're incredibly lucky and have tuned them to the exact same pitch.

3. Change the tuning of one of the strings slightly so it's either closer or farther away in pitch from the second string. Pluck the strings again and listen for the wah-wah sound. It should be either faster or slower than the first time. It should be faster the further apart the two pitches are and slower the closer they are together. If you found an electric guitar, you can pluck both strings and then vary the tuning of one while both notes still sound through the amplifier. The kids can then hear a smooth transition from slow to rapid wah-wahs.

Howcome, huh?

The wah-wahs you hear are called **beats**. Beats occur whenever you play two notes that are similar in pitch. In **Dueling Speakers** you learned that when sound waves or similar notes combine, they can add to each other in varying amounts or even completely cancel each other out. The resulting loud and soft periods are heard as beats. When the pitches are almost identical, the time between loud and soft beats gets very long. At some point, the time is so long that the beats essentially disappear. Musicians use the disappearance of beats to tune instruments. The wah-wah pedal for a guitar creates beats (among other effects) by electronically altering the incoming notes.

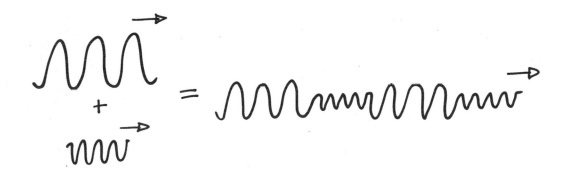

Sound Beats

Extensions

1. Challenge the kids to create beats with other musical instruments. Two people whistling can even create beats.

2. A guitar with all the strings tuned to the same note wouldn't make for a very interesting instrument. Find out exactly how guitarists and other musicians use beats to tune their instruments. Musicians with a tin ear can use fancy electronic equipment that relies on beats for tuning.

Sound Beats

Name_____

1. Explain what beats are.

    _____

    _____

2. What happens to the frequency of beats when two notes get further apart in pitch? What does "frequency" mean?

    _____

    _____

3. What happens to the frequency of beats when two notes get closer in pitch?

    _____

    _____

Symphonic Metal

You get some pretty tones and surprising effects just by whacking a metal rod.

Materials

1 metal rod, approximately ⅛" to 1" diameter, and 1' to 3' long
1 metal object such as a screwdriver, wrench, or butter knife

Set Up

Head to the hardware store to get the metal rod. You can use just about any kind, including threaded rods and re-bar. The folks at the hardware store will know what re-bar is.

The Zing!

1. Hold the center of the metal rod tightly between your thumb and forefinger.

2. Strike the end of the rod with a hard, metal object. Hear a nice, bell-like tone? If not, move your fingers a little bit and try again. You'll know when your fingers are in the right place, 'cuz when they're not, you'll hear a thud instead of a ring.

3. Now hold the rod about one-fourth of the way along its length and rap it again. Adjust your fingers until you get a ring instead of a thud. The ring you get this time should be a higher pitch than the one you got before.

4. Experiment with holding the rod in different places and hitting it. You should be able to get lots of different notes.

Howcome, huh?

When you whack a metal rod, you get many different kinds of sound waves traveling up and down the rod. A wave and its reflection add and cancel each other just like the other sound waves we've dealt with. For a rod of a given length, thickness, and composition, you end up with a pattern of waves known as standing waves. Standing waves are composed of parts that move a lot (these are called antinodes) and parts that don't move much at all (these are called nodes). The diagrams below represent some of the standing waves you create in a metal rod that's free to vibrate at any point (in other words, one that's not being held, even in the middle). These different vibrations combine to make a nice sound. When you hold the rod tightly in the middle, you prevent some of the standing waves in the above diagram from occurring. That's because you're keeping the rod

Symphonic Metal

from vibrating in the middle. In particular, the standing waves shown in the second and fourth drawings get "damped out," so you hear a combination of the notes corresponding to the first and third diagrams. When you hold the rod one-fourth of the way along its length, you eliminate all but the third drawing, and you hear this note. Holding it one-third of the way along the length gives you a note corresponding to the second drawing.

Extensions

1. Invite a good guitarist or violinist to the class to demonstrate the very pretty sounds you can get by gently placing your fingers at strategic places on the strings. The musician will know this as playing "overtones." Bring out your metal rod and have a jam session.

2. Get lots of metal rods of the same length and thickness and have the kids compose songs and play them. As long as the rods are almost identical, the different overtones will blend nicely and make some pretty music.

3. Go to the toy store and get one of those long, corrugated, plastic tubes that whirl in a circle to create sounds. Twirling the tube at different speeds results in different standing wave patterns and different notes. For extra credit, figure out how the **Bernoulli Effect** comes into play.

Symphonic Metal

Name_____

In the space below, explain why holding a metal rod in different places and whacking it creates different notes.

Metal Beats

Twirl a metal rod and create your own beat.

Materials

1 metal rod (as in previous activity)
1 piece of thread
1 metal object for whacking the rod

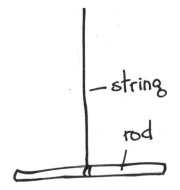

Set Up

Tie the thread around the metal rod at its center so the rod balances horizontally when you hold it by the thread. Make sure the thread is tight so the rod doesn't slip out and bang you on the toe while you're trying to impress the class.

The Zing!

1. Hold the rod by the thread and whack it near the middle. You should get a really nice tone — better than in the previous zinger because your fingers aren't damping out any of the notes.

2. Grab the rod so it stops ringing. Then whack the rod on the end so it spins horizontally. Ask the kids what they hear. If you've done the activity **Sound Beats**, they should recognize the beats this time.

3. Whack the rod so it spins at different rates. Notice any difference in the frequency of the beats.

Howcome, huh?

Providing you did the zinger **Sound Beats**, you know that beats are caused by two notes that are close in pitch. Since you only have one metal rod to bang, where do the two notes come from? It has to do with something called the **Doppler Effect**. When a car or train comes toward you, you hear it making a higher pitched sound than when it's moving away from you. You can really notice this as the object moves by you and the sound you hear shifts from the high to the low pitch. (Think of what race cars sound like as they whiz by and you've got the idea.)

The sound is higher when the object comes toward you because the sound waves get bunched up in front of the object. This causes the waves to hit you faster and results in a higher frequency. The sound is lower when the object moves away from you because the sound waves get spread apart, causing a lower frequency.

The faster the object moves, the more bunched up and spread apart the sound waves get, and the greater the difference between the high pitch (coming toward you) and the low pitch (going away from you).

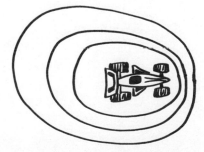

Metal Beats

When you hit the metal rod, it just sits there singing away at one frequency. But when you spin it, part of the rod is moving toward you and part is moving away from you. Because of the Doppler Effect, the part coming toward you sounds higher than normal and the part moving away sounds lower than normal. So you have two different pitches, and these pitches add together to create beats. The faster the rod spins, the greater the difference between the two pitches and the faster the beats.

Extensions

1. Take a field trip to the local music store or have the kids check out the large amplifiers the next time they're at a rock concert. Some amplifiers have a part that spins at different speeds. It's creating beats just as you did with the metal rod.

2. Do a little research into the Doppler Effect for light waves. Find out how it is used to figure out the speed at which the Sun and planets are rotating (it's a lot like the spinning rod). The Doppler Effect also provides evidence that the Universe is expanding.

Metal Beats

Name_____

1. When you spin a whacked metal rod, you hear beats. Where do the two notes that cause the beats come from?

   _____

   _____

2. Give an everyday example of the Doppler Effect.

   _____

   _____

3. Invent an instrument that uses the Doppler Effect and beats to create interesting sounds. Draw your instrument in the space below.

   _____

   _____

Polarized Light

Use polarized filters to make some interesting effects with an overhead projector.

Materials
  3 polarized filters or two pair of cheap polarized sunglasses
  1 overhead projector.

Set Up
If you're using sunglasses, remove three of the plastic shades from the frames. Now you know why "cheap" was specified. Polarized sunglasses are made of the exact same material as polarized filters. Set up the overhead, practice a bit, and you're ready.

The Zing!
1. Turn on the overhead and place one of the filters on the viewing area. Ask the class to describe what they see. The filter obviously filters out some light, creating an area darker than the surroundings.

2. Place a second filter on top of the first. Slowly rotate the second filter. It will go from shaded (some light getting through) to completely black (no light getting through), back to shaded. Ooohs and aaaahs.

3. Rotate the filters so no light gets through. Ask the kids what will happen when you add a third filter to the two. Go ahead and place the third filter on top of the other two; the area will remain black.

4. Now insert the third filter in between the other two and rotate it until light gets through the three-filter combination. Make sure the third filter covers only a portion of the other two filters so the kids can see that no light at all is getting through the two-filter combination. How can adding a filter cause more light to go through instead of blocking more out?

Howcome, huh?
In most everyday situations, light acts like waves on a string. In case you haven't watched waves on a string in a while, get a piece of string as long as your room and tie one end to something. Hold the other end fairly tight and stretch the string out. Move your end up and down rapidly and notice the waves that travel up and down the string. If you have a Slinky™, substitute that for the string and you'll get even better looking waves.

There are two main differences between light waves and waves on a string. The first is that light is composed of changing electric and magnetic fields, different than the simple motion of a string. The second difference is that ordinary, un-polarized light waves vibrate in more than one direction. They actually vibrate in a 360-degree circle. Check out the drawing.

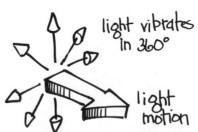

light vibrates in 360°

light motion

Polarized Light

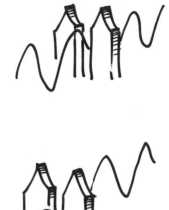

Back to simple waves on a string. Suppose you're going to send string waves through a narrow picket fence. No problem if the motion of the string is in the same direction as the openings in the fence.

But it's definitely a problem if the motion of the string is perpendicular (at right angles) to the openings in the fence. Such waves won't make it through.

Now suppose you have a string with wave motion that travels in all sorts of directions. You are trying to send all those waves through a picket fence. The fence will select out only those waves with motion, or a component of motion, in one direction — the direction that's the same as that of the openings in the fence.

Polarized filters do exactly the same thing to light waves. The waves come into the filter oscillating (moving up and down, back and forth, etc.) in all directions. The only light waves that make it through the filter, though, are those oscillating in certain directions. Only half the light gets through. After passing through the filter, the light is "polarized," which means that it's now oscillating in only one direction.

Before explaining the zinger, you might like to know that polarized filters are, in fact, a whole lot like picket fences. They're made of long chains of molecules that are pretty much all lined up in one direction. Although it's not like the light actually is passing through long narrow spaces between molecules, that's essentially the result.

Now back to picket fences. What if you try to send string waves through two picket fences that have their long, narrow spaces at right angles to each other? The first fence selects out only waves traveling in a certain direction. Since the second fence is at right angles to the first, the waves emerging from the first fence won't get through the second fence.

Likewise, when you rotate two polarized filters so their long chain molecules are at right angles, no light can get through. Rotate them just a little from this right angle orientation, and some light gets through. The most light gets through when the long chain molecules in each filter are oriented in the same direction.

Finally, you have two filters at right angles (no light getting through) and you slip a third in between. Voilà! Light gets through. To understand this, just look at two of the filters at a time. The bottom one and the middle one aren't at right angles, so some light gets through the middle filter. The middle one and the top one aren't at right angles, so some of the light that gets through the middle one also makes it through the top one. If you don't buy that explanation, then . . . well, it's magic!

Polarized Light

Extensions

1. Place a piece of plastic food wrap between two polarized filters that are at right angles (no light getting through). No surprises, right? Now stretch the plastic wrap and it'll act just like a polarized filter, mainly because it now <u>is</u> a polarized filter. When you stretch plastic wrap, you pull the long-chain plastic molecules into a picket fence orientation.

2. Take a polarized filter outside and look through it at the sky, in a direction about 45 degrees away from the sun's position in the sky. Rotate the filter and you'll find that indirect sunlight is polarized. The scattering of sunlight by the atmosphere, in addition to polarizing the light, makes the sky blue.

3. Take the kids outside and show them how polarized filters reduce glare. That is, look through a polarized filter at sunlight reflecting off water or metal. Reflected light is polarized, so a filter removes some of the reflected light. Obviously, the orientation of the polarized filters is important in the design of sunglasses.

Polarized Light

Name_____

In the space below, draw at least four pictures that show what happens when you rotate one polarized filter on top of another.

Polarized Colors

Make some very cool colors by combining common items and polarized filters.

Materials
- 2 polarized filters or 1 pair of polarized sunglasses
- 1 clear plastic or glass cup
- 1 overhead projector
- clear corn syrup
- plastic food wrap

Set Up
Check out the previous activity for an explanation of how to use the sunglasses if you don't have polarized filters. Then fill the cup with clear corn syrup and wheel the overhead projector into place.

The Zing!
1. Just in case you thought the overhead was for looks only, turn it on. Place one of the filters flat on the overhead.

2. Place the cup of clear corn syrup on top of the filter.

3. Hold the second filter just above the liquid and rotate it in a horizontal plane. You should get a steady progression of the different colors of the rainbow. Oooooh!

4. Remove the cup of corn syrup. Crumple up a sheet of plastic wrap and place it on top of the bottom filter.

5. Hold the second filter over the plastic wrap and slowly rotate it in a horizontal plane. Many different colors to please the little ones' eyes.

Howcome, huh?
Both plastic wrap and corn syrup rotate the direction of polarization of the incoming light. Because you have a filter on the bottom, the light from the projector hitting either material is polarized in a certain direction. The direction of polarization of the light has already rotated by the time it leaves the corn syrup or plastic wrap. The pretty colors are created because different colors of light rotate the direction of polarization different amounts.

Let's apply this first to the corn syrup. Each color of light passing through the syrup has its direction of polarization rotated a definite amount. This amount depends on the color and on the thickness of the liquid. For example, a 2-inch layer of syrup will rotate through red light 45 degrees (approximately). The filter on top only lets the light that's partially or completely lined up with the direction of its long-chain molecules. When those molecules are at 45 degrees relative to those of the bottom filter, you'll see mainly red light.

Polarized Colors

Rotate the top filter a bit farther, and the dominant color you'll see is the one that has its direction of polarization rotated a bit more than 45 degrees. As you continue rotating, you reach the rotation that corresponds to the different colors that make up white light.

The explanation for the crumpled plastic wrap is essentially the same, except for the fact that crumpling creates many different thicknesses of plastic wrap through which the light may travel. Different thicknesses rotate a given color different amounts, so you end up with a multi-colored version of what happens with the corn syrup.

Extensions

Experiment with different liquids and clear solids to find out which rotate the direction of polarization of light. Any liquid with sugar in it or any clear plastic material will work for starters.

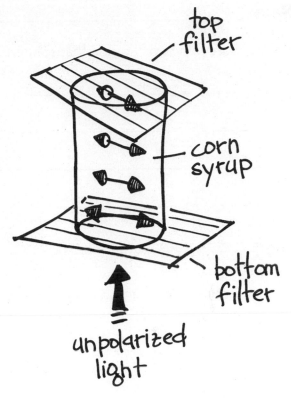

top filter

corn syrup

bottom filter

unpolarized light

Polarized Color

Name_____

1. List 5 materials or liquids that rotate the direction of polarization of light.

_____

_____

_____

_____

_____

2. List 5 materials or liquids that won't rotate the direction of polarization of light.

_____

_____

_____

_____

_____

Water Magnifier

Use water droplets to magnify and reduce a visual image.

Materials

1 beaker
1 eyedropper
1 jar of petroleum jelly
1 large washer
1 microscope slide
water

Set Up

This is a very easy zinger to set up. Just fill the beaker with water and go see if anybody brought any goodies to eat.

The Zing!

vaseline

1. Dip your finger into the petroleum jelly and coat one side of the washer thoroughly. Depending on the kind of washer you have, you might find that one side is flatter than the other. In that case, grease up the flat side. The jelly acts as a gasket and prevents the water from leaking out from under the washer.

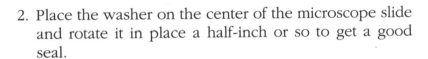

water

washer

slide

2. Place the washer on the center of the microscope slide and rotate it in place a half-inch or so to get a good seal.

3. Tell the students you are going to magnify the words in a book using a drop of water. Fill the eyedropper full and add drops to the washer until it looks like a small molehill; use the drawing at the center right as a guide. Place the lens over the words that you would like to magnify and you'll find that they are much larger. Move the slide up and down over the words and observe the change in images.

4. Now tell the students that you are going to shrink the words that you just magnified. Using the eyedropper, draw the water from the inside of the washer until the water level makes a small dip. Place the lens over the words and you'll find that they are smaller than the actual print. Again, move the slide up and down over the words and observe the change in images.

Water Magnifier

Howcome, huh?

When light travels from one material to another, such as from air to water or vice versa, it bends. This bending is called **refraction**. There are exact rules for how much and in which direction light bends when it goes from one material to another, but since you already have enough rules added to your complicated day, simplicity is best. In the first case, the light coming from the words on the page bends inward. This makes the light rays from those words seem to be coming from a different place. As you can see from the diagram, this makes the words look larger than actual size.

image appears larger

In the second case, the light bends outward, causing the words to look smaller than they actually are. Again, refer to the incredibly clear diagram.

Eyeglasses and contact lenses work on this principle. People who are nearsighted or farsighted have eye lenses that either bend light too much or too little. Putting another lens (the glasses or the contacts) in front of the eye bends the light in the opposite way, correcting for the error in refraction. Laser surgery that corrects vision problems changes the shape of the cornea, much like you can change the shape of the water droplet by adding or removing water. Any change in shape changes the amount of magnification or reduction, and can correct the way your eye focuses.

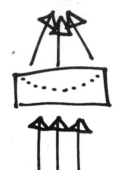

image appears smaller

Extensions

1. Round up enough materials for each student or group and this makes a great class activity. It is also a perfect time to have the local optometrist or ophthalmologist come on out and whip out some lenses, eye charts, and laser surgery techniques to impress the students.

2. There are plastic lenses you can slap onto the back window of your van or motor home that give you a 180° view. These lenses are known as **Fresnel** (fr˘e - nell) **lenses**, not surprisingly named after a guy named Fresnel. They work on the same basic principle of refraction of light, and if you want to go crazy, try and figure out how all those ridges create a 180 degree view. At any rate, the kids will think it's cool.

3. Contact the local junior high or high school physics teacher and see if he or she has an optics bench to bring by for the students to see. These usually consist of a fancy meter stick, some lens holders, lenses, and mirrors to demonstrate the properties of light and lenses.

4. Check the offerings at the local museum and see if they have anything that deals with lenses.

Water Magnifier

Name_____

In the circles below draw pictures of the letters that you looked at through the magnifying lens. The circles on the left are the objects that have been magnified and the ones on the right are the ones that have been reduced.

magnified   **reduced**

Electricity and Magnetism

What you have here is a brief collection of low-maintenance, highly entertaining electrical and magnetic zingers. For the record, you can't get a serious electrical shock no matter how much you abuse or misuse the equipment.

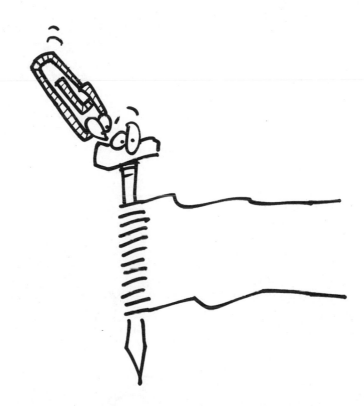

Dancing Papers

Bring a charged balloon near a beaker filled with pieces of paper. Watch the papers dance.

Materials
- 1 balloon
- 1 beaker or large glass
- 1 pair of scissors
- 1 sheet of paper
- hair

Set Up

No special prep for this one. If you can, choose a dry day. Any experiment that has to do with static electricity tends to work better when there is lower humidity. This is bad news for you folks in the Pacific Northwest, but so goes life.

The Zing!

1. Cut the paper into small pieces, roughly a quarter-inch square, and put them under the beaker (or glass). You can substitute pencil shavings.

2. Blow the balloon up and tie it off. Rub the balloon on a wool sweater, a student's head, or a passing dog.

3. Bring the balloon near the beaker. The pieces of paper will be attracted to the balloon and dance and fall as you move the balloon around the beaker. If you are having trouble with the experiment, check to make sure that the inside of the glass is dry. If you are still hitting a glitch, try a thinner glass or wait until a drier day.

Zero to Einstein in 60™   146

Dancing Papers

Howcome, huh?

Everything is made of atoms and these atoms have both negative (electrons) and positive (protons) charges. When you rubbed the sweater or your hair, you actually rubbed some of the electrons (negative charges) off onto the balloon. The balloon now has this huge negative charge bouncing all over it. What a molecular nuisance!

The balloon has a negative charge. Whoa. This charge is placed near some sweet innocent pieces of paper. The big, bad, huge old ugly negative charge on the balloon scares the tiny, weeny, little electrons to the far sides of each of the atoms in the paper. The result is that each atom in the paper is slightly polarized. The entire piece of paper is still electrically neutral, but the side closest to the balloon has a net positive charge and the side farthest from the balloon has a net negative charge. (See the illustration in the upper right hand corner). Because electrical forces depend on distance, the closer attraction of opposite charges overcomes the farther repulsions of the like charges. The polarized condition of the atoms in the paper cause it to head towards the negatively-charged balloon.

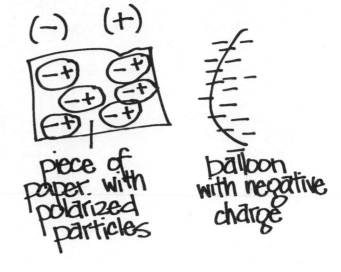

Extensions

1. For an adaptation, rub the balloon again and stick it to the wall. Decorate your whole room. (Remember this is science, not fooling around; principals traditionally need an education in this field.)

2. In fact, if you have tenure, take the whole class into the principal's office when he or she is not there and decorate the entire office with balloons.

3. Or better yet, use those little packing peanuts made out of Styrofoam in place of balloons. They work great.

Dancing Papers

Name_____

1. Did the balloon attract the pieces of paper when it was uncharged?

   _____

2. What material was used to charge the balloon?

   _____

3. List any other materials you can use to charge the balloon.

   _____

   _____

Sticky Balloons

Stick a balloon to the wall after rubbing it vigorously on someone's head.

Materials
    1 balloon
    1 source of electrons (hair, wool, stray cat)

Set Up
No prep needed. Just pull out your balloon and let 'er rip.

The Zing!
1. Inflate the balloon and tie off the end. Hold it to the wall and release it. If everything is going according to plan, the balloon will tumble to the floor.

2. Wool is the best material to use for generating a charge on the balloon. Rub the balloon very rapidly on the chosen surface for five seconds and then hold the balloon to the wall. It should stick. If the balloon doesn't stick, it might be because there's too much humidity in the air. The less the better. You might also try other surfaces to see if you can collect more electrons off them.

Howcome, huh?
All materials are made up of particles called atoms. Each of these atoms has a positively-charged middle called a nucleus and little whirly gigs zipping around that middle, called electrons, which have a negative charge. (It is important to note that the electrons do not act like little planets as they bop around the outside of the atom; they are more like a fly circling around your dinner table. You know that it will be in a general vicinity but the exact location is hard to pin down.)

When you rub the balloon, you collect these electrons from the surface of the wool or the hair, and the balloon now has an excess of electrons on its surface, giving it a huge negative charge. As you hold the balloon near the wall, the negative charge on the balloon pushes the negative electrons towards the far side of each atom. The atoms in the wall that are closer to the balloon are thus polarized, with one end being slightly positive and one end being slightly negative. Overall, the positive charges in the wall are closer to the balloon than are the negative charges. Because electrical forces depend on distance, this makes the attraction between unlike charges stronger than the repulsion between like charges. The balloon sticks.

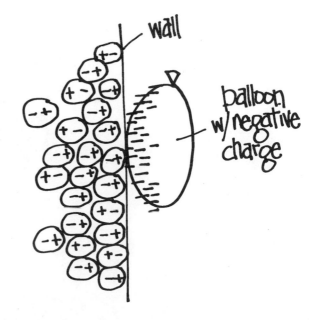

Sticky Balloons

Extensions

Find other materials that will stick to the wall when you rub them on your hair, a cat, etc. For starters, try Styrofoam and felt.

Sticky Balloons

Zinger #47

Name_____

List five objects in the room to which the balloon will stick. Then list five objects that repel the balloon.

Will Stick	**Will Not Stick**
1. _____	1. _____
2. _____	2. _____
3. _____	3. _____
4. _____	4. _____
5. _____	5. _____

Floating Magnets

Surround a pencil with two circular magnets. One floats above the other.

Materials
    2 circular magnets
    1 pencil

Set Up
Just grab the materials and practice your presentation.

The Zing!
1. Announce that you have discovered how to make things float in mid-air. Hold the pencil upright and slide the magnets over the eraser. Before you do this, make sure you know which side of the magnets are opposites and slide them so they stick. Time for a puzzled look.

2. Announce that you will have to have a word with the offending and sometimes slothful magnet on the top. Remove this magnet and have a very quiet, stern visit with the magnet, making sure that you flip it over. This time it will float above the other magnet. Peace is preserved in the land and wild applause breaks out.

Howcome, huh?
All magnets have a north and a south end, or **pole**. Opposite poles attract one another and like poles repel. When you slid the magnet on top of the other magnet the second time it floated because two like poles repelled each other.

Extensions
1. Try variations on the arrangement. If you have access to more magnets, try mixing three, four, and five magnets.

2. Experiment with magnetic objects around the room to see if you can get them to levitate in the air.

3. Have the kids try this without the pencil. Just about impossible!

4. Speaking of impossible, see if you can get a bar magnet to float above another bar magnet.

Floating Magnets

Zinger #48

Name_____

Test five objects in each category listed below. After you have tested all 20 objects, write a rule that explains why magnets like to stick to some things and not to others.

Glass

1. _____

2. _____

3. _____

4. _____

5. _____

Wood

1. _____

2. _____

3. _____

4. _____

5. _____

Metal

1. _____

2. _____

3. _____

4. _____

5. _____

Plastic

1. _____

2. _____

3. _____

4. _____

5. _____

My Rule:

_____

_____

_____

Electromagnet

Wrap a nail with wire, connect the wire to a battery, and create an electromagnet.

Materials

1 nail, 16-penny (uncoated)
1 box of straight pins
1 battery, 6-V
1 bar magnet
copper wire (or bell wire), #22 insulated, about 4 feet

Set Up

Strip about one inch of insulation off the ends of the wire.

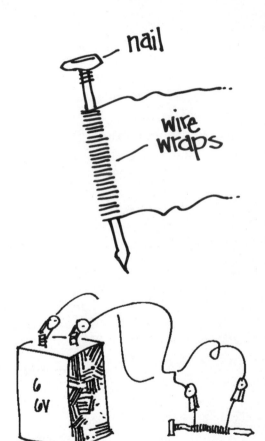

The Zing!

1. Empty the pins out on the table. Ask a student volunteer to gently touch the nail to the pile of pins and ask the class if the nail is acting like a magnet. Not yet!

2. Using the copper wire, begin at the top of the nail and wrap the wire around it. As you begin, leave a 6" tail at the top and wrap the wire tightly, trying not to leave spaces. When you get near the end of the nail, loop the wire under the previous loop and cut the wire so you have another 6" tail.

3. Hook each tail to one of the terminals on the battery and have your student volunteer touch the electromagnet to the pile of pins. The nail will now pick up several pins.

4. Unhook the battery and demonstrate that once the nail has been magnetized it will remain magnetized after the battery is removed. If you wish to demagnetize the nail, all you have to do is take it out of the wire casing and toss it on a hard surface once or twice. This will discombobulate the iron atoms and they will lose their magnetic pull.

Howcome, huh?

When you hook up a wire to a battery so you have a complete circuit (no breaks), the electrons in the wire move. The movement of electrons (or any other charged particles, for that matter) is known as an electrical current. Any wire that carries an electrical current will exert a force on magnets. In particular, a long, current-carrying coil like the one you made in this activity acts just like a bar magnet, with a north and south pole.

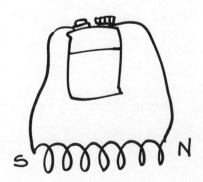

Electromagnet

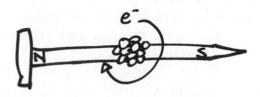

Even without the nail, your coil of wire is magnetic whenever it's hooked up to the battery. If you experiment a bit, you'll find that the coil with the nail inside is a much stronger electromagnet. To understand why, you have to take a microscopic look at the inside of a nail. Each atom in the nail has a magnetic moment (sounds romantic, huh?), which is just a fancy way of saying that each atom acts like a tiny magnet. One way of looking at it is to think of the electrons in an atom circling around the nucleus, creating an electrical current (electrons in motion) which, like a large-scale current-carrying wire, creates a magnet.

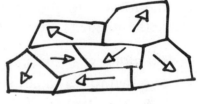

The magnetic moments of individual atoms combine together to form magnetic domains inside the nail. Each domain also acts like a tiny magnet. In your garden-variety nail, the magnetic domains are randomly oriented, pointing in all sorts of directions. Because of these different directions, the magnetic effects of the domains cancel each other out.

magnetic domains in an un-magnetized nail

Magnets tend to line up with each other. When you bring a strong magnet close to a nail, the magnetic domains inside the nail tend to line up with that magnet. (It's not as if they just rotate around, but it's something like that.) Since a current-carrying coil of wire is a magnet, the domains in the nail line up with the north-south direction of the coil of wire wrapped around the nail.

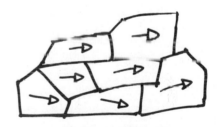

Now that the domains in the nail are pointing in pretty much the same direction, the nail itself is a magnet. When you unhooked the battery, the magnetic domains in the nail reverted back to their original directions. You should have noticed, however, that the nail remained slightly magnetic

magnetic domains in a magnetized nail

even after the battery was disconnected. That's because the magnetic domains don't immediately return to their original directions. When you whacked the nail on a hard surface, you jumbled up all the magnetic domains inside, and the nail was demagnetized. Heat can cause this to happen, as well, so don't let your refrigerator magnets hang out on the oven.

Permanent magnets are just materials in which the magnetic domains stay lined up even without an electrical current or another magnet around. They can still be demagnetized when dropped, which is why that old set of bar magnets that have been clanging around your school for 20 years just don't seem to work very well.

Electromagnet

Extensions

1. Have the students experiment in order to figure out the relationship between the number of wraps on the nail and the strength of the magnet (as measured by the number of pins that it picks up).

2. Ask the students why neatness is important when constructing the electromagnet. Encourage them to create two electromagnets using the same number of wraps. One magnet should be very neat and tight, the other should be piggy. Same number of wraps.

3. Ask the local auto shop guy to come on down and show the students how starters work and where the magnets are located.

4. Try the Radio Shack™ dude. Speakers are driven by magnets and electromagnets. Invite him or her down to tear apart some extra electronic equipment.

5. Finally, in an effort to integrate the arts and sciences, rent some Road Runner™ cartoons and show those to the class. Inevitably Wile E. Coyote™ gets himself into a predicament with large electromagnets. Save this extension for the day you're getting evaluated and you're sure to hit the jackpot.

Electromagnet

Zinger #49

Name_____

Two experiments have been outlined for you below. In the first one, you will vary the number of wraps going around the nail. In the second experiment, you are to make one set of wraps as neatly as possible and the other as messy as you want. Report your findings in the spaces below.

Experiment 1

Number of Wraps Around Nail	Number of Items Picked Up
10	
20	
30	
40	
50	

Experiment 2

Types of Wrap Around Nail	Number of Items Picked Up
Very messy & Loose	
Sort of Messy	
Neat	
Very Neat	

Magnetic Nails

Magnetize a 16-penny nail with a bar magnet and pick up some pins.

Materials

1   16-penny nail (uncoated)
1   bar magnet
1   pile of straight pins (not too big)

Set Up

Occasionally nails don't know that they are not supposed to demonstrate magnetic qualities. Test the nail prior to the students coming into the room and see if it will pick up any pins. It shouldn't. If you have an over-enthusiastic nail, all you have to do to demagnetize it is throw it on the ground (hard surfaces like concrete work best) a couple of times. This will rearrange the magnetic domains in the nail enough so they won't attract the pins.

One other thing you will have to watch out for is a coating on the nails. These nails are used in outdoor construction because they resist corrosion. These cannot be magnetized very well at all and should be avoided unless you wish to inflict minor amounts of brain damage on the students.

The Zing!

1. Hold the nail close to the pile of pins and demonstrate that the nail is not magnetic.

2. Hold the nail in one hand and, using the bar magnet, brush the nail in one direction. It's like brushing your hair — you always brush down, never up. The important thing is to brush the same way and not rub the nail back and forth.

3. Touch the nail to the pile of pins again and several should stick to it. Cool!

4. Take the nail and give it a quick toss to a hard surface. Demonstrate the loss of magnetism by trying to retrieve pins again. If it picks up pins, try one more toss.

Howcome, huh?

If you could look inside a piece of iron you would see that it consists of many small magnetic "domains" which are usually arranged in a random fashion. Each of these domains consists of many atoms and can be moved, stretched, and directed to create a magnetic field. It helps to imagine a field of wheat. The whole field represents the whole piece of iron, and the individual stalks of wheat represent the domains. On a calm day the wheat just sort of hangs around going any direction it wants. If the wind picks up the wheat starts to move together, bending and wiggling in the same

Magnetic Nails

direction. When the bar magnet is repeatedly brushed over the surface of the nail, the domains are aimed in one direction. Recall that magnets tend to line up with one another. This coordination of the magnetic domains creates a weak magnetic effect and can then attract and hold small objects that contain iron, such as straight pins, paper clips, and iron filings.

Extensions

1. The more you brush the nail the stronger it gets, to a certain point. Let the kids experiment and figure this out for themselves. In fact, have them create a data table and record the results of their tests as they increase the number of rubs. The data table might look something like this.

# Rubs	# Pins

2. Another interesting experiment is to brush two nails and hold them together to see whether or not they pick up double the number of pins.

3. See if the size of the nail influences the number of pins that it picks up. Do bigger nails have a greater ability to become magnetized or is it the quality of the nail and not the size?

4. Bouncing the nails off the ground is one way to demagnetize them. What happens if you carefully heat the nail in a candle? What if you placed it in the freezer?

5. Can a magnetized nail magnetize another nail?

Magnetic Nails

Name_____

Design an experiment to test the magnetic powers of your nail. How does the number of times you brush the nail with the magnet affect its power to pick up pins?

Materials

_____

_____

_____

Procedure

_____

_____

_____

Data

_____

_____

_____

Conclusion

_____

_____

_____

_____

Herculean Magnet

Increase magnetic force by adding a couple of washers to a circular magnet.

Materials

circular magnets (one will work but more is better),
metal washers, 2 per magnet
   (Washers should have about the same diameter as the magnets.)

Set Up

Nada.

The Zing!

1. If you only have one circular magnet, have a volunteer come up and demonstrate for the class. If you have more, form groups and pass them out to the class. For what follows, you'll need one volunteer.

2. Give the magnet to the kid and ask him or her to hold the magnet near something metal that is attracted to the magnet (part of a desk, scissors, whatever).

3. Ask the volunteer to judge which part of the magnet is the strongest — the flat sides or the round edge. It should be pretty clear that the flat sides are the strongest. Head into your favorite discussion of magnet poles if you feel like it (the flat sides are the north and south poles of a circular magnet).

4. Have the kid slap a metal washer on both sides of the magnet so it looks like the drawing.

5. Ask the volunteer to again determine which part of the magnet, the side or the edge, is strongest. The edge wins by a landslide, and it's also much stronger than the sides were before.

Howcome, huh?

The poles of a magnet are much stronger than the rest of the magnet. That's why the flat sides (without washers) are much stronger than the rounded edge. Adding metal washers messes this up; in particular, they direct the magnetic force off to the side. The edge of the circular magnet now has its original magnetic force plus the re-directed force from the poles. Hence, a Herculean magnet.

Extensions

1. If you have more than one circular magnet, check the edge magnet strength of a bunch of them in a stack compared to that same stack with washers in between the magnets.

2. Have the kids experiment with metal pieces and different kinds of magnets (bar magnets, horseshoe magnets, etc.) and try to alter magnetic strengths in as many ways as possible.

Herculean Magnet

Name_____

1. Which part of a circular magnet is strongest, the flat sides or the round edge?

   _____

   _____

2. Explain how adding washers to the magnet changes your answer to the first question.

   _____

   _____

   _____

3. Invent a carnival ride that uses what you've learned in this activity. Draw a picture of the ride below.

3-D Magnetic Field

Create a hairy but cool spectacle out of iron filings and a cylindrical magnet.

Materials
- 1 cow magnet (obtain from a science catalog or a local feed store)
- 1 plastic bag, 1 quart or smaller
- 1 small hot-dog-shaped balloon
- 2 ounces of iron filings (approximately)

Set Up
Go get a cow magnet. Feed or farm supply stores are the cheaper option, but science catalogs might be more convenient. If you go to the feed store, act like you know what a cow magnet is. Perhaps reading the explanation to this activity will help.

Iron filings are also available from science catalogs, but your students will get the same result by dragging the magnets through sand and dirt to collect all the iron filings they need.

The Zing!
1. Pour the iron filings into the plastic bag. Drop the cow magnet inside the uninflated balloon. The whole purpose of the balloon is to keep the iron filings from coming in direct contact with the magnet, so the size of the balloon isn't critical. Once the magnet is inside, wrap the balloon tightly around it.

2. Dip the balloon-wrapped cow magnet into the bag of iron filings and shake it around a bit.

3. Pull the magnet out and you should have a hairy-looking representation of the magnetic field of the cow magnet. Encourage the kids to touch (gently). It's a visual-tactile experience that's extraordinary.

4. To retrieve an iron-filing-free magnet, place it back in the bag and roll the balloon down around the outside.

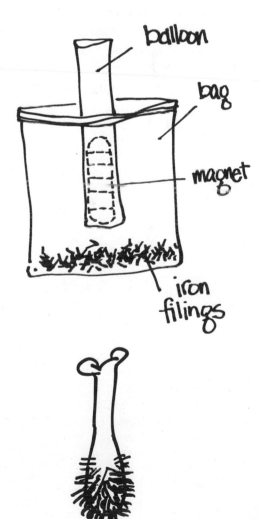

3-D Magnetic Field

Howcome, huh?

Iron filings become magnetized when in the presence of a strong magnet. (Refer to the two previous activities.) They then tend to line up with the magnetic field of that strong magnet. A magnetic field is an invisible pattern around a magnet that tells what kind and how strong of a magnetic force that magnet will exert on other objects. By placing a bar magnet under a sheet of paper and sprinkling iron filings on top of the paper, you get a two-dimensional picture of the magnetic field of that magnet. The current zinger simply extends that to three dimensions.

As for the purpose of cow magnets: Ranchers sometimes have a difficult time rounding up their cattle. In such cases, they have each cow swallow a cow magnet. The rancher then turns on a powerful electromagnet, which pulls the cows to the desired location.

If you bought that explanation, there are a number of bridges someone would like to sell you. In fact, cattle sometimes eat, along with their healthy diet, various metal objects such as cans and barbed wire. If this metal were to travel through a typical cow's digestive tract, you would have a dead cow. Cow magnets are inserted into the first of a cow's four (!) stomachs, where they trap any eaten metal and prevent it from going on through the digestive tract. Really. Honest.

Extensions

Use iron filings to map out the magnetic fields of all sorts of different magnets.

3-D Magnetic Field

Name_____

In the space below, draw a picture of what the cow magnet looked like when your teacher took it out of the bag of iron filings.

You can physically or chemically alter the structure of things, which is what this section is all about. You'll begin with basic, mundane fire; move to pretty colors; and finish with shattered tennis balls.

Out of Oxygen

Demonstrate that oxygen is one of fire's most essential ingredients.

Materials
   1 beaker or clear glass jar
   1 candle
      matches

Set Up
Just pull stuff out, set it on the table, and chase down your missing stapler.

The Zing!
1. Hold up the beaker and ask the students what's inside it. There are only two options here: air and possibly your fingers.

2. Light the candle and place it under the beaker. The flame will gradually diminish and go out. Not very spectacular, but it does illustrate a point very well.

Howcome, huh?
As the candle burns, much of the oxygen in the beaker combines with other atoms and is no longer a gas. When a great deal of the oxygen is gone from the air, the flame dies. To carry this a step further, in order for anything to burn it must have three things: fuel (the wax and wick), heat (supplied by the match), and oxygen (found in the air surrounding the candle under the beaker). If you remove any one of the three, the fire is extinguished.

Extensions
1. Use larger containers and time how long the flame lasts.

2. You can conduct a health-related experiment by having the students breathe into the container after they have held their breath for varying amounts of time (10, 20, 30, 40 seconds, etc.). The longer they hold their breath, the more carbon dioxide their bodies will dump into their lungs. When this is transferred to the container, the candle stays lit for a shorter period of time.

3. Get hold of a bottle of oxygen and fill the beaker with pure gas. When you invert the container you will notice that the flame burns brighter.

Out of Oxygen

Name_____

Number the containers from one (the container where the candle will go out fastest) to four (the container where the flame will burn the longest).

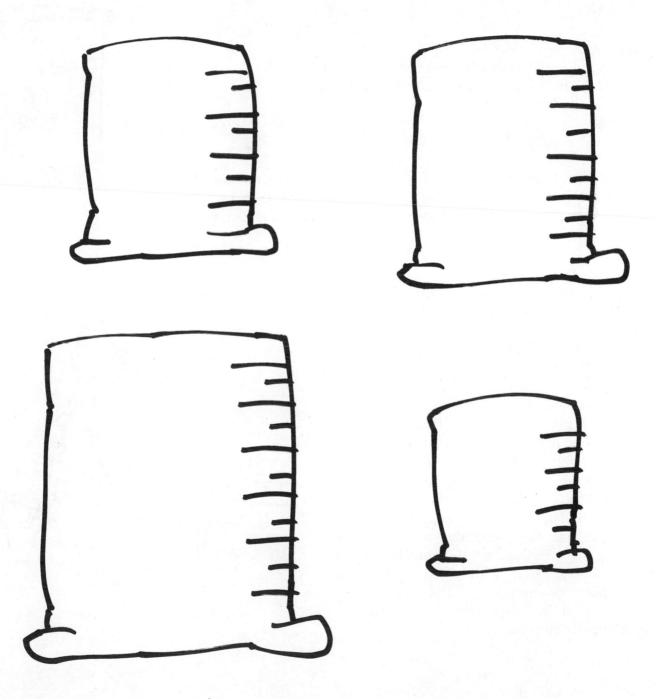

Oatmeal Extinguisher

Extinguish a candle by thumping on the back of an oatmeal box.

Materials
1 candle
1 empty oatmeal cylinder
   matches

Set Up
Find an empty oatmeal cylinder and cut a hole in the middle of the top of the lid. It should be about a half-inch in diameter.

The Zing!
1. Light the candle and set it on a table.

2. Hold the cylinder in front of the candle, approximately six inches away. Make sure the hole is pointing toward the flame, and give the back of the cylinder a good thump. The flame will go out.

Howcome, huh?
The air inside the container is forced forward by the thump on the back. When the air reaches the opening, it is compressed into a stream and extinguishes the flame. Which brings up to the question of how blowing air across a flame causes it to go out. After all, when you blow on a fire in a fireplace, the fire burns brighter! One of the things a fire needs to keep burning is heat. When you blow across a flame, you're providing plenty of oxygen, and the wax fuel (see next activity) is there. However, you also remove the burning gases. You're removing the heat necessary to keep the fire burning, and it goes out. A fireplace fire continues to burn only because you're dealing with a much greater source of heat to begin with. When you blow into a fireplace or campfire, you simply provide more oxygen for the tremendous amount of heat and fuel already present.

Oatmeal Extinguisher

Extensions

1. Experiment to see how far away you can get and still put the candle out.

2. Vary the opening of the diameter to see if there is any influence on the demonstration.

3. Have the kids determine whether the size of the container makes any difference. Using the data gathered from extension Steps 1 and 2, ask the kids to create the optimal flame extinguisher based on the diameter of the opening and the size of the container.

Oatmeal Extinguisher

Name_____

Test the oatmeal extinguisher to see how far away you can get and still put out the flame.

Distance (cm)	Lit	Extinguished
1		
5		
10		
15		
20		
25		
30		
35		
40		
45		
50		

 ©1989 Rev 1999 The Wild Goose Company WG-3005

Flame Pinchers

Pinch off a candle's fuel supply with a pair of scissors.

Materials
    1 long tapered candle
    1 pair of tweezers
       matches

Set Up
Have the materials ready to go and you'll be fine.

The Zing!
1. Light the candle and allow it to burn for a few seconds.

2. Take the tweezers and pinch the wick just below the flame.
   It should gradually go out.

Howcome, huh?
This zinger is great for demonstrating that one of the three things necessary to make a fire is **fuel**. If you've been paying attention, the previous two "Zingers" addressed the need for oxygen and heat. A quick inspection of the candle just below the wick will reveal that the wax is liquefied. Once it is in this liquid state, it travels up the wick where it burns. By pinching the wick, you are cutting off the fuel supply and the flame goes out.

Extensions
1. To demonstrate that it is the wax that allows the candle to burn for so long and not the wick, light a length of string on fire. The students will see that it burns very quickly.

2. A great magic trick can be used to follow up on this experiment. When you extinguish the candle flame, you will notice that there is a stream of smoke that curls up from the wick for several seconds. This smoke is actually the gaseous form of wax and is highly flammable.

   As you extinguish the candle, light another match. As soon as the flame goes out, bring the lit match over to the stream of hot wax gas. The gas is flammable, and as soon as you touch the flame to it the stream will reignite and the flame will appear to jump down to the wick and relight the candle.

   If you have trouble lighting the wick it may be one of three things. You have waited too long to relight the candle, you may be too far from the wick, or there may be a draft in the room causing the smoke to disperse too quickly. Close the door and hold the match close to the wick as fast as you can.

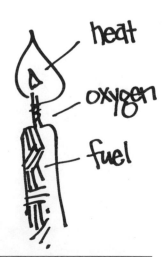

Flame Pinchers

Name_____

Write a poem about extinguishing a candle using a pair of tweezers. Don't use any words longer than five letters. You can make it sad if you want; after all, the candle will be extinguished.

Liquids to Solid

Blend two liquids together to form a solid.

Materials
   2 test tubes
      calcium chloride
      sodium carbonate

Set Up
First, take note that this activity requires sodium carbonate and not sodium bicarbonate, which is baking soda. Once you have the chemicals, they're most likely in a powder or crystal form. What that means is you'll have to make up your own solutions. Not too tough, though. Just stir and dissolve as much of each chemical as you can in warm water (use separate containers)! If you end up with chemical in the bottom that won't dissolve, just pour off the liquid and use that.

The Zing!
1. Fill one-third of one test tube with calcium chloride solution and hand it to your student volunteer.

2. Fill one-third of the second test tube with sodium carbonate solution and also hand that to your student volunteer.

3. Present both tubes to the class and ask them to predict what will happen when the two solutions are mixed together. Pour the contents of one test tube into the other test tube. Instant solid.

Howcome, huh?
When you mix the two solutions together, the calcium chloride and sodium carbonate react to form new chemicals — sodium chloride (table salt) and calcium carbonate. Both of these are solids and fall out of solution. The fancy name for the fallout is a **precipitate**.

Extensions
You can do this same demo substituting potassium carbonate for sodium carbonate. Within limits, the kids can try to create precipitates by combining other solutions. CAUTION: NEVER COMBINE ANYTHING CONTAINING AMMONIA AND BLEACH. Instant nerve gas.

Calcium
(Ca)
Chloride
(Cl₂)
Potassium
(K)
Carbonate
(CO₃)

$$CaCl_2 + K_2CO_3 \rightarrow 2KCl + CaCO_3$$

Liquids to Solid

Name_____

Describe the contents of each of the test tubes.

Calcium chloride:

_____

_____

Sodium carbonate:

_____

_____

The tube they were mixed in:

_____

_____

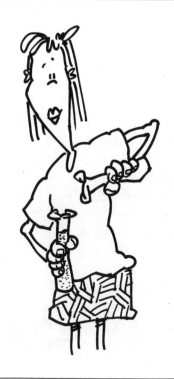

Rusting Steel

Push a balloon into a test tube filled with steel wool and a few drops of water.

Materials
    1 beaker or glass jar
    1 piece of steel wool
    1 test tube
      water

Set Up
There is no real preparation needed for this experiment.

The Zing!
1. Place the steel wool inside the test tube. Wet the steel wool thoroughly with water and dump off the excess water.

2. Fill one-fourth of the beaker or jar with water. Invert the test tube and place it in the beaker so the mouth of the test tube is submerged.

3. Check the set-up every half-hour or so. After a while, you'll notice that the water from the beaker is rising up inside the test tube. Leave things long enough, and the water level in the test tube will rise higher than the water level in the beaker.

Howcome, huh?
When steel comes in contact with water, it rusts, which is an everyday way of saying that oxygen from the air combines with the steel. Since the test tube is sealed off from the outside air by being submerged in the beaker of water, the oxygen for the rusting process comes from the air inside the test tube. So the rusting steel wool removes oxygen from the inside of the tube. When you remove air molecules from a confined space, you reduce the air pressure there (see the air pressure section of the book). With a lower air pressure inside the tube, the outside air pressure pushes the water in the beaker up inside the tube. And as mentioned before, the water isn't sucked up inside the tube because science never sucks.

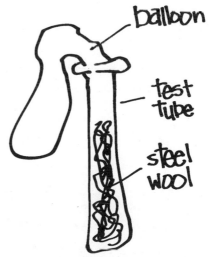

balloon

test tube

steel wool

Extensions
One adaptation is to set up a **control**. A control is just a scientist's way of making sure to test a single idea in an experiment rather than two or three. Make a second tube just like the first but don't add the water. This way you can check the importance of the water.

Rusting Steel

Name_____

1. Is the test tube an open or closed system? (Can anything get into or out of the tube?)

   _____

   _____

2. List three things that are inside the test tube.

   _____

   _____

3. What happens to the water level in the jar or beaker as the steel rusts?

   _____

   _____

4. What causes this reaction?

   _____

   _____

Rainbow in a Bottle

Add acetic acid to water and a chemical indicator called bromothymol blue. Watch the liquid change from blue to yellow.

Materials

1 pipette
1 tall, clear glass or test tube
1 spoon (straight from the kitchen drawer)
  acetic acid
  bromothymol blue
  1M sodium hydroxide solution
  water

Set Up

The bromothymol blue solution can be made by dissolving 0.16 grams of bromothymol blue in 2 liters of water.

The Zing!

1. In the glass, mix one part bromothymol blue with two parts water.

2. Using the first pipette, add drops of acetic acid one at a time until a color change occurs. Careful observation reveals a change from blue, to green, to yellow.

3. Now use the second pipette and slowly add drops of sodium hydroxide to the glass. Stir the contents of the glass every several drops and note the color changes. If you are careful as you add the base, you will see that the colors will exactly reverse. The solution will turn to green and then blue again.

Howcome, huh?

Bromothymol Blue is an acid base indicator that operates in a pH range of 6.0 to 7.6. When it's in a basic solution it's a blue color, when neutral it's green, and when slightly acidic it's yellow. As the acetic acid is added to the glass, it acidifies the water slowly. The more acetic acid that is added to the water, the lower the pH gets. When it reaches a neutral level it turns green, and when it becomes acidic it turns yellow.

When sodium hydroxide, a base, is added a couple of drops at a time, the solution becomes more basic, the pH rises, and the colors reverse themselves.

Rainbow in a Bottle

Extensions

1. Have the kids use a straw to blow bubbles in to the bromothymol blue solution. Carbon dioxide from they're lungs combines with water to form carbonic acid, which causes a color change.

2. Have the kids add various household liquids such as orange juice, liquid soap, and bleach to the bromothymol blue. CAUTION: NEVER MIX AMMONIA AND BLEACH.

Starch Balls

Mix and play with balls of cornstarch.

Materials

cornstarch
water

Set Up

Read through the instructions, get out your apron, and practice this one; it may take some time but it's fun.

The Zing!

1. Add a pile of cornstarch that is about as big as a silver dollar to the palm of your hand. Add water, a little bit at a time, until you have an almost runny paste. This is the part that takes the practice.

2. Moving in small circles, mix with your other hand. If the paste is the right consistency it will turn from a runny, watery goop into a solid paste ball. The paste ball exists as long as the pressure is applied. If you stop rubbing the paste it returns to the liquid form.

   *If the paste starts to dry out add a dab (a very small amount) of water. Too much and you have glop that is not manageable.

Howcome, huh?

When the cornstarch mixes with the water it does not mix completely. There are still several spaces that the water hasn't penetrated. When the mixture is under the pressure of the rubbing hand, the water is forced into these spaces temporarily; once the pressure is removed it goes back out and you have glop. Fun, huh?

Extensions

1. Mix up a huge tub of this stuff and play with it. If you move your hand slowly through it, you won't have trouble. Move quickly and you're stuck; much like quicksand.

2. Have the kids design a spaceship that can land on a planet of this stuff without sinking in.

Starch Balls

Name_____

1. What happened when you rolled the starch mixture in your hand?

    _____

    _____

2. What happened when you stopped putting pressure on the starch?

    _____

    _____

3. Suggest two possible reasons why this happened.

    _____

    _____

Frozen Tennis Ball

Remove a frozen tennis ball from a bucket, toss it on the ground, and watch it shatter!

Materials

    3 tennis balls
    2 buckets
      dry ice
      regular ice
      heavy gloves

Set Up

Thirty minutes before you perform this zinger, place one of the tennis balls in a bucket of dry ice. Be sure that the ball is completely immersed in the stuff. Use heavy gloves as you do this. Don't let the students get a hold of the dry ice. Cold burns are free for the taking and trips to the hospital are not out of the realm of possibility. Pack a second ball in a bucket of regular ice.

The Zing!

1. Show the room temperature tennis ball to your students. Bounce it around the room and talk about elasticity and the ability of the ball to absorb and rebound from the shock of hitting other objects, be it a wall, the ground, or a racket.

2. Remove the tennis ball from the regular ice and toss it on the ground. It will bounce a little bit, but not as much as the room temperature one.

3. Now take the ball that has been sitting in the dry ice and toss it on the ground. It will shatter into a bunch of pieces. Hmmmmm.

Howcome, huh?

The particles in the third ball have been cooled to nearly minus 109 degrees Fahrenheit. At this temperature, materials lose their elasticity and become very brittle. When the ball is tossed onto the ground, it can no longer absorb the shock and breaks into pieces.

Extensions

You can cool virtually anything. Rubber bands are a lot of fun as are oranges, racquetball balls, and anything else that your imagination can come up with. Actually, the hot dogs might be the most fun of all.

Frozen Tennis Ball

Name_____

Draw pictures of the really cold tennis ball before and after the demonstration.

Why did the tennis ball shatter?

_____

_____

_____

_____

NOTES

NOTES

NOTES

NOTES

NOTES

NOTES